Supply Chain Management in African Agriculture

Abdul-Razak Alhassan ·
Mamudu Abunga Akudugu

Supply Chain Management in African Agriculture

Innovative Approaches to Commodity Value Chains

palgrave
macmillan

Abdul-Razak Alhassan
Department of Strategy, Management and Marketing
Birmingham City University
Birmingham, UK

Mamudu Abunga Akudugu
Institute for Interdisciplinary Research and Consulting Services
University for Development Studies
Tamale, Ghana

ISBN 978-3-030-54208-5 ISBN 978-3-030-54209-2 (eBook)
https://doi.org/10.1007/978-3-030-54209-2

This Palgrave Macmillan imprint is published by the registered company Springer Nature Switzerland AG
The registered company address is: Gewerbestrasse 11, 6330 Cham, Switzerland

Preface

For decades, the transformation agenda of Africa has been a mirage, largely because of the weak performance in the mainstay of African economies—agriculture. The history of Africa is well tied to its vast natural resources including fertile lands that support agricultural production. It has the potential to produce enough to feed itself and export to other continents. Unfortunately, it continues to rely on food imports and food aid from Europe and the Americas. The main reason for this is the weak agricultural commodity value chains. This is the common story across Africa and Ghana is no exception. Indeed, during the 2019/2020 cropping season, rice farmers in Ghana lamented over lack of market for their produce, a situation exacerbated by lack of storage facilities. And yet, the country spends hundreds of millions of dollars annually importing rice. To halt and reverse the imports dependency syndrome of African countries, there must be a focus on the transformation of the agricultural sector through commodity value chain development.

A general Google search on agricultural commodity value chains in Africa indicates that there is no much material on it. Where the material exists, the content does not go beyond theoretical abstractions to indicate the specific business models that could work for individual commodity value chains. So, the content of this book would benefit policymakers, policy implementers, development practitioners, agri-entrepreneurs, researchers and all those who have interests in the transformation of African agriculture among others. It will be an excellent reference material for students of agriculture, agribusiness, agricultural economics, rural development and so on. This book no doubt provides practical business models for the transformation of African agriculture through commodity value chain development.

The aim of this book is to provide holistic and critical information on the supply chain management in African agriculture using the tomato value chain to illustrate what needs to be done to change the narrative of Africa being a continent that is generally a net importer of food and food products despite its vast agricultural potential. The overarching objective is to propose business models that can transform African agriculture supply and value chains.

Aylesbury, UK — Abdul-Razak Alhassan
Tamale, Ghana — Mamudu Abunga Akudugu

Contents

Acronyms

AGOA	African Growth and Opportunity Act
AMSEC	Agricultural Mechanization Services Centre
CEPS	Customs Excise and Preventive Service
CET	Common External Tariff
CRI	Crops Research Institute
CSIR	Council for Scientific and Industrial Research
CWGs	Chain Working Groups
DFIs	Designated Financial Institutions
DPs	Development Partners
ECOWAS	Economic Community of West African States
EDAIF	Export Trade, Agricultural and Industrial Development Fund
EPA	Economic Partnership Agreement
ETLS	ECOWAS Trade Liberalization Scheme
EU	European Union
FAO	Food and Agriculture Organization
FASDEP	Food and Agriculture Sector Development Policy
FBOs	Farmer-Based Organizations
FDA	Food and Drugs Authority
FDI	Foreign Direct Investment
FOHCREC	Forest and Horticultural Crops Research Centre
FTA	Free Trade Agreements
GAIDA	Ghana Agricultural Input Dealers Association
GATRA	Ghana Agricultural Transformation Agenda
GEPA	Ghana Export Promotion Authority
GIDA	Ghana Irrigation Development Authority
GLSS	Ghana Living Standards Survey

GSA	Ghana Standards Authority
GSGDA	Ghana Shared Growth and Development Agenda
HYV	High Yielding Varieties
ICOUR	Irrigation Company of Upper Region
MDAs	Ministries, Agencies and Departments
MESTI	Ministry of Environment, Science, Technology and Innovation
METASSIP	Medium Term Agricultural Sector Support and Investment Plan
MOFA	Ministry of Food and Agriculture
MOFEP	Ministry of Finance and Economic Planning
MSMEs	Micro, Small and Medium Enterprises
MTDP	Medium Term Development Plan
NDPC	National Development Planning Commission
NGOs	Non-Governmental Organizations
SARI	Savannah Agricultural Research Institute
SSM	Special Safeguard Mechanism
SWOT	Strengths, Weaknesses, Opportunities and Threats
UDS	University for Development Studies
UN	United Nations
WTO	World Trade Organization

List of Figures

List of Tables

CHAPTER 1

Background Issues

Abstract The majority of people in developing countries, particularly those in Africa live in rural areas where agriculture remains the main source of employment because of limited livelihood alternatives. This means that agriculture is at the centre of efforts to attain the Sustainable Development Goals (SDGs) in Africa. Thus, African governments and their development partners invest substantial amounts in the sector. Despite these investments, the sector continues to be sluggish and characterized by weak commodity value chains. The focus continues to be on production at the farm level with less emphasis on value addition. This chapter lays the foundation on innovative approaches to promoting commodity value chains necessary for transforming the sector through effective and efficient supply and commodity value chain management.

Keywords Africa · Agriculture · Supply chain · Value chain · Agriculture clusters · Technology

1.1 Introduction

Agriculture is and will continue to be the centrepiece of the economies of most developing countries, especially those in Africa. This assertion is in line with the empirical literature that the economies of countries

A.-R. Alhassan and M. Abunga Akudugu,
Supply Chain Management in African Agriculture,
https://doi.org/10.1007/978-3-030-54209-2_1

in Sub-Saharan Africa (SSA) are largely dependent on agriculture, which employs over 60% of the labour force and accounts for about 35% of Gross National Product (GNP) as well as 40% of foreign exchange earnings (Fulginiti et al., 2004). Thus agriculture is a major sector, which provides livelihoods, especially in rural Africa. This is particularly so because majority of the population in Africa resides in rural areas, with an estimated 70% of the workforce being engaged in agriculture for survival (Fan et al., 2009). Diao et al. (2007) argued that growth in agriculture is one of the most effective ways of reducing poverty, improving rural livelihoods and promoting economic growth in Africa. However, the inherent risks associated with the sector make populations that hitherto depended on agriculture diversify their livelihoods into non-farm activities such as petty trading (Akudugu, 2011) and small-scale mining, which is fast becoming an important livelihood activity in rural Africa including Ghana (Banchirigah, 2006, 2008; Hilson, 2002).

The central role of the agricultural sector in the pursuance of sustainable socio-economic development in SSA with the overall objective of poverty reduction and improvement in livelihoods can only be attained with improvement in productivity performance in the sector. Enhanced crop and agricultural productivity is thus seen to be critical for the overall socio-economic and politico-cultural well-being of the people of SSA, especially those in the rural areas. A number of multi-country studies (Examples include Block, 1994; Dewbre & Borot de Battisti, 2008; FAO, 2000; Frisvold & Ingram, 1995; Fulginiti et al., 2004; Lusigi & Thirtle, 1997; Rao & Coelli, 1998; Suhariyanto et al., 2001; Thirtle et al., 1995) have therefore been conducted over the years to look at how crop and general agricultural productivity performance across countries in SSA can be enhanced for sustainable livelihoods and socio-economic development. Despite the fact that these studies are different in terms of geographical coverage and timings, the central point is that there has been little variation in their findings. According to Fulginiti et al. (2004), almost all studies conducted in the region over the past decade report productivity gains in the 1960s, productivity losses or stagnations in the 1970s and recovery to productivity gains in the 1980s and early 1990s with varying degrees of influences on livelihoods. To sustain the positive gains in productivity, there is the need to harness the vast natural resource base of the region to ensure that agriculture continues to play its expected roles in the regional development agenda.

African governments have generally shown their commitment to improving rural livelihoods by putting the agricultural sector at the top of their development priorities (Dewbre & Borot de Battisti, 2008). That notwithstanding, the sector is said to be underperforming (InterAcademy Council, 2004) and this has left many people who largely depend on agriculture for their livelihoods in abject poverty. This is consistent with Dewbre and Borot de Battisti (2008) who reported that large percentages of people who depend on farming for a living in SSA are still in poverty. According to them, there are widening income gaps between farmers and non-farmers and rural populations continue to suffer from food insecurity, malnutrition and social exclusion.

In the book entitled, *Building Competitiveness in Africa's Agriculture: A Guide to Value Chain Concepts and Applications*, Webber and Labaste (2010) noted that there has been a strong resurgence of interest by actors in the African agriculture and agribusiness sectors in promoting value chains as an approach that can help design interventions geared to add value, lower transaction costs, diversify rural economies and contribute to increasing rural household incomes in SSA countries. The authors further argued that enhancing value chain competitiveness is increasingly recognized as an effective approach to generating growth and reducing the rural poverty prevalent in the region. This has been the long held view of practitioners who often argue that there is the need to look at agriculture differently—not just as a means of survival, but as smaller or larger commercial businesses linked to domestic and global markets—and of the need to identify and tap into new sources of potential growth and value addition in the sector. This might only be possible with effective and efficient supply chain management in African agriculture. This book is therefore particularly designed for those who want to know more about innovative approaches to commodity value chains for sustainable and inclusive development in Africa. It focuses on commodity value chain-based approaches, and how to use them in ways that can contribute to sound operational decisions and results for enterprise and industry development, as well as for policy-making with respect to doing business, stimulating investment and enhancing trade in the context of African agriculture.

1.2 Methodological Approach to Commodity Value Chain Profiling

There are varied methodological approaches for profiling value chains and these range from complicated and strict quantitative modelling based on positivist epistemological stance to a more flexible approach based on constructivism. It is important to note that commodity value chain profiling must be done in a manner that captures the interests, objectives, goals and aspirations of all actors. The methodological approach that does this well in our considered opinion is the participatory approach based on pragmatism. This methodological approach uses participatory tools and techniques to capture the key issues at each segment of the value chain within the socio-economic, politico-cultural, technological and climatic circumstances that actors operate.

In using the participatory methodological approach to value chain profiling, the process is divided into phases. The first phase of the process focuses on reviewing the empirical and grey literature on the specific commodity value chain that is being profiled. The second phase is a Community-based stakeholder's consultative process, which allows us to map the inflows and outflows generated by actors along the value chain. This approach follows the Coady International Institute (2012) leaky bucket concept, which includes producer-led market mapping, market exploration, identification of key actors, identification of opportunities and bottlenecks and listing of potential interventions/activities.

The fact that the participatory methodological approach is inspired by the pragmatists' epistemological stance means that quantitative and qualitative tools and techniques could be freely drawn for commodity value chain profiling. These may include the use of surveys, key informant interviews and stakeholder/focus group discussions. The key benefits of using a participatory methodology for the profiling include farmers' empowerment and increased understanding of the commodity value chain among actors. This methodology creates the awareness among actors along the commodity value chain on the need to reflect on their individual/group situations within the value chain; providing the opportunity for actors to identify activities that could be taken up at individual, group, community and national levels to strengthen the weaker segments and/or sustain the stronger segments of the value chain, with or without any external support and resources. The methodology is made of six main steps briefly described in the following subsections.

1.2.1 Cluster and Products Identification (Step I)

The first step is to among other things identify the clusters and products of interest (e.g. fresh tomatoes, onions or cabbage for hotels). The identification of clusters and products must be based on empirical literature and stakeholder consultations. Commodity leaky buckets are then used to capture the inflows and outflows at each segment or actor level along the value chain. This involves highlighting outflows, which capture the expenses incurred in producing the commodity of interest (e.g. buying seeds, fertilizers, chemicals, fuel for irrigating the land, labour, etc.) and inflows, which captures the money earned through the sales of output to different market channels. This is done through individual interviews and focus group discussions with stakeholders and key industry actors. The product 'leaky bucket' concept is then used to investigate the markets for inputs, services and outputs in the next step.

1.2.2 Market Map for Inputs, Services and Outputs (Step II)

The second step is to develop a market map for inputs, services and outputs using results from the first step. This process involves finding out where the inputs for growing the specific commodity come from and where the outputs are sold. To do this, stakeholders, particularly farmers are asked to indicate the sources of each of the inputs and services, picking one at a time from the product 'leaky bucket' in Step I. The output markets where farmers sell their produce are also mapped. The markets in most of Africa may include roadside retailers who buy directly from the farmers and sell to customers on busy national highways, open-air markets, grocery stores, supermarkets, processing companies, institutions, etc. The outcome from this step is fed into the next step of identifying the different market channels and actors.

1.2.3 Market Channels and Actors (Step III)

The third step is to visit different markets and actors identified in the market map in Step II. The visits are to create the platform for stakeholder interactions and this enables the profiler, researcher or practitioner to interact with key industry players and to assess the roles and levels of influence of each actor in the value chain. This step is critical as it creates the opportunity for actors along the commodity value chain to appreciate the fact that their actions either positively or negatively influence the overall effectiveness of the value chain. The outcome from this step feeds into the next step of developing a value chain map that is practically relevant to the key actors operating within the different segments, effective, efficient, sustainable, impactful and coherent.

1.2.4 *Development of Value Chain Map (Step IV)*

The fourth step is to develop a value chain map by finalizing the cluster maps and business models suitable for the particular commodity/industry. The value chain map captures all the major actors in the commodity value chain and the existing support actors and their functions and relationships. The outcome here is used for the next step of undertaking the strengths, weaknesses, opportunities and threats (SWOT) analyses.

1.2.5 *Identification of Strengths, Weaknesses, Opportunities and Threats (Step V)*

Based on the mapping exercises, field investigations and interviews, the strengths, weaknesses, opportunities and threats (SWOT) emerging at the different segments along the commodity value chain are documented and analysed.

1.2.6 *Upgrading Needs (Step VI)*

The activities needed to upgrade the value chain are identified here. Thus the appropriate activities needed to sustain the strengths, overcome the weaknesses, take advantage of the opportunities and minimize the influence of the threats are identified. Thus, the appropriate activities that rely on the strengths to deal with the weaknesses and opportunities to overcome the threats are listed at this stage. The essence is to be able to come out with well-tailored activities that will make the entire commodity value chain economically viable, politically stable, gender sensitive, environmentally sustainable, culturally agreeable and social acceptable to all actors. To do this, the value chain is divided into four main segments or arenas as:

1. Inputs segment/arena
2. Services segment/arena
3. Production segment/arena
4. Output marketing segment/arena.

1.3 Conceptual Descriptions and Benefits of Clustering

1.3.1 The Concept of Agriculture Clusters

From a socio-economic perspective, an agriculture cluster is a system that brings together economic and social stakeholders who participate in coordinated activities that add value to any agricultural produce, from its production until it reaches the consumer. A cluster includes producers, providers of inputs and services as well as processing, industrialization, transportation, logistics and other support services such as financing (Garcia-Winder et al., 2009). It must be noted that the process of segmentations and aggregation of value is neither lineal nor egalitarian. On the contrary, the arrangements among the different segments of an agriculture cluster more closely resemble a 'web' of non-lineal relationships that can be highly inequitable, as certain stakeholders with strong negotiation, management, economic or political power could dominate and extend their influence over the weaker, less organized players who have less influence in the decision-making processes. These relationships can also exist on numerous levels with the value of products, goods and services frequently altered.

From an analytical point of view, agriculture cluster is a way of understanding the relationships or links among the stakeholders of agriculture and the rural milieu; from the supply of inputs and primary production to the delivery of the product to the final consumer where the relations established may be of a contractual or commercial nature. From the operational perspective, agriculture cluster may be seen as an institutional arrangement for strategic planning, policy management, dialogue and consensus-building among stakeholders or as a social contract where the government, the private sector and civil society establish short- and long-term commitments for the comprehensive development of a particular agri-food value chain (e.g. tomato value chain).

From a business point of view, agriculture cluster can be used as a tool to regulate relations and arrangements between private organizations, improving the terms of transaction, the business results and relations among the stakeholders. In this sense, the agriculture cluster has a clearly defined place in time and space, which responds to specific market conditions or processes.

1.3.2 The Concept of Agri-Food Value Chain Clusters

Agri-food value chain clusters are defined as a geographic agglomeration of competing and related businesses, where there is evidence of improved performance such as a growth and profitability (Kuah, 2002). The most important point to make is that cluster and value chain concepts are not mutually exclusive given that a cluster forms part of the value chain. Agri-food value chain clusters are important and could be used as tools for analyses and mechanisms to facilitate dialogue and promote commitment among actors so as to define public policies to improve competitiveness. Agri-food value chain clusters are therefore used to generate methodologies for the analyses of chains and to support the creation of chain organizations as well as monitor their management.

The use of agri-food value chain clusters, as tools to regulate relations among actors must be based on a common denominator—the search for greater transparency in business transactions and a balanced dialogue among the actors involved in these processes. When agri-food value chain clusters are used in this context, and all the actors are included, their use as a business tool facilitates the inclusion of smaller players in business transactions. This generates opportunities to improve the incomes of actors operating in the weakest segments of the chain and encourages large commercial firms to implement policies of social and environmental responsibility. This implies that agri-food value chain clusters may be regarded as instruments for achieving greater equity and participation. For this reason, both the public and private-sector links of the chain must be willing to share timely and transparent information, with the assurance and confidence that it will be used to improve their overall competitiveness and not to unilaterally favour some of the actors to the disadvantage of others.

1.3.3 The Concept of Commodity Value Chain Clusters

Commodity value chain clusters are closely linked to agriculture and agri-food clusters and can be understood to encompass the entire flow of inputs and outputs that enables farm enterprises to realize the value of their unique capital base by meeting the needs of final consumers. The unique capital base of commodity value chain cluster actors consists of natural capital (land and water), physical capital (equipment, livestock and crop inventories), specialized human capital and financial capital owned and employed by the different players along the value chain in their operations. The commodity value chain cluster is therefore a geographically proximate set of interconnected enterprises and associated institutions linked by commonalities and complementarities where co-location of the various stakeholders accelerates knowledge sharing and development of new products and services in the industry. Simply put, a commodity value chain cluster is an interconnected group of actors in the industry who are found within a given agroecological area the emergence and growth of which have implications for the productive sectors of the economy and geographic area within which it is situated.

1.3.4 Benefits of Clustering in the Tomato Value Chain

There are numerous benefits associated with the development of commodity value chain clusters. Agribusinesses and related companies as well as individual actors within a commodity value chain cluster will benefit from economic advantages of engaging within a cluster. In other words, individuals, agribusinesses, institutions and the economy of the geographical area will generally benefit from a commodity value chain cluster development through increased employment opportunities with government generating more tax revenues resulting in better service provisions and higher standards of living. By operating in a cluster, agribusinesses and individual actors along the different segments of the value chain will find it easier to share information, tacit capabilities, specialized human capital and other common-access intangible resources than when they are located at a distance from one another.

Clustering holds important implications both for the agriculture sector and the wider economy of production areas generally. By clustering the value chain, industry players who are part of the value chain and geographic areas within which such clusters arise stand to benefit immensely from cluster activities. The clustering will ensure that firms and individuals operating along the commodity value chain have the opportunity to collectively introduce new technologies that will lead to lower production costs, higher revenues and profits as well as improved industry performance. When this happens, the competitive positions of key industry players at the local and global levels will be enhanced, thereby making their products competitive at the national and international levels.

Indeed, getting actors along the local commodity industry to be effective and competitive is key. This is possible through clustering, which will enable local producers, input suppliers, service providers, traders, transporters and processors to be proximal to one another for easy sharing of resources and collective investments in the deployment of cutting edge production, transport and market-related technologies. It is therefore envisioned that clustering will lead to overall holistic development of cluster areas because as agriculture and related enterprises within clusters succeed and grow, both direct and indirect employment opportunities are created. For instance, a well functioning commodity value chain cluster in a given geographical area will not only be good for agriculture enterprises and agribusinesses but also those in the other productive sectors of the economy—services, industry, manufacturing, etc. This is because the pool of actors within the cluster creates demand for the goods and services of actors in the other productive sectors thereby resulting in a rippling effect on the local and national economy. The creation of commodity value chain clusters will result in input suppliers, service providers, processors and marketers locating into the cluster areas resulting in decreased coordination costs and creation of opportunities for more complex vertical and horizontal linkages among firms and individual actors. The fact that enterprises within a cluster have greater incentive to undertake collective investment decisions will lead to decreases in production costs and risks.

In summary, the development of agriculture clusters and commodity value chain clusters in particular is critical for accelerated transformation of the agricultural sector in Ghana. As noted by Porter (1998, 2008), clusters often extend downstream to customers and laterally to manufacturers of complementary products and to companies in industries related by skills, technologies or common inputs. The ultimate benefit is that as clusters grow, there will be increases in employment opportunities, sales and tax revenues and this will make local government authorities better equipped to efficiently provide essential services to the citizenry resulting in higher standards of living.

1.4 Ecosystems, Climate Change and Technology for Commodity Value Chains

1.4.1 Ecosystems Services and Commodity Value Chain Development

The well-being of society and natural capital are closely linked (Burkhard & Maes, 2017; Diaz et al., 2018; Tasser et al., 2020). This is because the livelihoods of people are very dependent on the goods and services derived from nature. It is therefore no surprise that the attention globally in recent years has increasingly focused on the importance of ecosystem services (ES). Several initiatives have therefore emerged in developed and developing countries to promote the protection and conservation of natural resources following the Millennium Ecosystem Assessment in 2005. One such important initiatives has been the Inter-governmental Science-Policy Platform on Biodiversity and Ecosystem Services (IPBES, 2019) multilevel assessment.

These initiatives have become relevant since biodiversity at both global and local levels are increasingly being threatened which will directly and indirectly affect livelihoods. Biodiversity loss changes the functioning of ecosystems, reducing their ability to provide essential goods and services to society (Costanza et al., 2017; Huge et al., 2020). The importance of ecosystems to livelihoods in the provision of regulatory services such as climate regulation, flood prevention, erosion control, water purification and air; provisioning services such as food and water, transportation; supporting services such as primary productivity, nutrient recycling; as well as cultural services such as educational, recreational, aesthetic services, are pivotal to livelihoods and general well-being of people.

Well developed and functioning agricultural commodity value chains are key in driving and mobilizing action for the protection and conservation of natural resources and providing education on the importance of ecosystems. This is because land use for agricultural and other purposes, is regarded a major contributor of GHG emissions with some researchers suggesting that this may surpass even fossil fuels (World Economic Forum, 2019). Agriculture, food production, deforestation among other human activities are significant drivers of human-induced climate change which negatively affects ecosystems. Furthermore, land degradation caused by human activities also contributes significantly to the loss of ecosystems and biodiversity. This could in part be attributed to the effects of human-induced land degradation on pollinators for example that are significantly impacted but are very much required in agriculture given that three out of every four of the commercially grown leading crop types according to IPBES (2019), rely on pollination from animals for field and quality. The available evidence further suggests that food systems for example, account for 25% of global greenhouse-gas emissions, 70% of freshwater withdrawals and 60–70% of biodiversity loss whiles climate change threatens up to 25% of crop yields (World Economic Forum, 2019) which all destroy ecosystems.

The strained food systems that are putting pressures on ecosystems, are further expected to be exacerbated by the increasing global population which is estimated to reach 1000A0billion by 2050, an increment of over 30% from current population figures (World Economic Forum, 2019).

A transformative approach both rooted in technology and innovative ecosystems is pivotal in addressing this strain. Enabling environments consisting of policy frameworks, governance models, business models, well-developed agricultural value chains as well as investments according to the World Economic Forum (2019), are the innovative ecosystems required to address the current food system challenges.

1.4.2 Climate Change and Commodity Value Chain Development

The consequences of climate change is already being felt globally with the frequent than normal extreme weather, rising sea levels among many other changes. These weather changes are impacting agricultural commodity value chains since the agricultural industry globally particularly farming lands, are vulnerable to the effects of climate change. These climate effects present significant challenges as well as opportunities that will have to be managed. By shifting to a more sustainable future and taking advantage of the benefits this presents, the agricultural sector is also well positioned to lessen the impact of the risks of climate change.

The effects of climate change such as extreme weather effects including low rainfall and draughts, floods, heatwaves, erosion among others, will significantly impact land ecosystems and biodiversity which will eventually impact production and profitability of the agricultural sector. Furthermore, climate change has the potential to cause much longer-term problems such as desertification and shifting climate zones. Climate change also directly and indirectly impacts other sectors key to commodity value chains including transport, energy, infrastructure among others, which gets damaged as a result of the physical effects of climate change. This could potentially result in productivity and profitability losses impacting the livelihoods particularly of smallholder producers.

Well-developed agricultural commodity value chains will help the agricultural industry to both mitigate the risks of climate change and also take advantage of the resultant numerous opportunities. Identifying and mitigating against these climate change risks will require detailed commodity value chain analysis and the development of sound commodity value chains to ensure the development of a resilient agricultural sector equipped with appropriate adaptive measures.

Furthermore, mitigation measures against climate change risks through commodity value chain development, must provide for an integrated holistic approach in response involving all interrelated sectors such as water, ecological, environmental among others (Fulbright, 2019).

1.4.3 Technology and Commodity Value Chain Development

The relevance of technologies in addressing inefficiencies in markets as well as meet global market entry requirements cannot be underestimated. As technologies become readily available particularly in developing countries, they will inevitably continue to transform agricultural commodity value chains significantly. In recent years, several technologies have been developed and deployed that are profoundly transforming agricultural commodity value chains in both developed and developing countries. These technologies including the use of drones, blockchains, robots among others, although generally accepted as transformational in enhancing agricultural value chains, are not without challenges. Regulatory constraints, lack of training, funding constraints, scalability as well as trust concerns are among the challenges confronting technology use.

Addressing these challenges is key in encouraging technology use and promoting effective and efficient commodity value chains. The widespread adoption and use of existing technologies in agriculture will require not only building trust among potential users particularly smallholder farmers on the effectiveness of the technical solutions these technologies provide, but also convincing them of how safe these are particularly with regard to data protection or the security of their transactions using mobile technologies for example. Furthermore, technology must also be simple enough to comprehend by potential users if it is to be widely accepted especially in developing countries. Additionally, the widespread use and adoption of technology in agricultural value chains comes at a cost. However, it is only through such widespread use can the relevance of technology be appreciated and its impact in agriculture realized. Finding access to finance to promote the development of relevant technological solutions to pertinent issues within commodity value chains and also incentivizing potential users such as smallholder farmers who otherwise may be reluctant to forgo old practices to adopt new practices, may encourage the adoption and use of technology.

Fourth Industrial Revolution technologies including Artificial Intelligence (AIs), Internet of Things (IoT), advanced robotics and Precision Agriculture according to the World Economic Forum (2019), could help disrupt as well as shape the transformation of food systems. Unfortunately, although these technologies are increasingly becoming available, the rate at which these technologies are being developed are still behind technologies being developed for other sectors, largely because of the relative lack of technological investments in agriculture.

This notwithstanding, the fact is that existing technologies today, have the potential not only to provide cost-effective solutions in agricultural commodity value chains but also in ensuring that food systems are able to meet market requirements and demands. Technology adoption and deployment in commodity value chains evidence suggest, is impacting food systems by promoting transparency and accountability, reducing fraud, improving food safety and more importantly, traceability in food systems across supply chains (World Economic Forum, 2019) which has become requirements in securing entry into profitable markets such as the European Union (EU).

In Ghana, for example, a traceability system for fruit and vegetable growers in the country, implemented by the USAID's Trade and Investment Programme for Competitive Export Economy that runs between 2005 and 2009 with the goal of increasing exports to the EU and increase the participation of smallholders in these markets using GIS and barcode applications with GPS research, barcode scanners, networked computers and a wireless mobile network, found in one of their programme evaluations that half of their mango value chain respondents indicated that they had expanded their mango-growing areas resulting in other livelihoods improvements such as improved ability to make home improvements and educate their children (World Economic Forum, 2019).

1.5 Globalization and Commodity Value Chain Development

1.5.1 Multilateral Trading Systems and Commodity Value Chain Development

Ghana is engaged in a number of multilateral trade regimes and has been part of most of the trade rounds including the Uruguay and Doha Rounds. These trade rounds have the potential to reverse the marginalization of some developing and least developed countries through a development-oriented approach. In particular, the Doha Round reinvigorated a multilateral trading system that would deliver opportunities to all participants, including the smallest, weakest and most vulnerable economies. It is important for countries in Africa to take a dynamic approach to maximize returns from the new trading opportunities that come with the Doha Development Agenda to alleviate poverty and put their economies on the path of sustainable growth and inclusive development that leave no one behind.

With specific reference to agriculture, the Uruguay Round saw a number of African countries including Ghana, a founding member of the World Trade Organization (WTO) bounding all their agricultural tariffs. For instance, about 99% of Ghana's agricultural tariffs were bound at a ceiling level of 99% effective 2004. Lower bound rates of 40 and 50% were set for a few agricultural products to apply from 1995. However, Ghana maintains a standard maximum applied rate of 20% on all agricultural products, with lower rates of 0, 5 or 10% applying to a quarter of these products. Ghana does not maintain any export subsidies nor does it maintain any trade-distorting domestic support and these greatly affect the agricultural sector. This is particularly so as Ghana is not immune from the deleterious effects of export subsidies and trade-distorting domestic support of some WTO members.

Other areas of the agriculture negotiations of special interest to Africa include the reduction or elimination of tariff peaks and tariff escalation on products of export interest to the continent, the self-designation of Special Products by developing countries to be treated flexibly in the negotiations, and a Special Safeguard Mechanism (SSM) to deal with import surges. Preference erosion is also a matter of concern to Africa.

1.5.2 Regional Integration and Commodity Value Chain Development

Regional economic cooperation and integration have serious implications for commodity value chain development. For example, member countries of the Economic Community for West African States (ECOWAS) protocols commit themselves to integration programmes and activities in the area of trade by taking actions on the following, all of which have implications for commodity value chains beyond countries:

- **Free Movement of Persons**: Member countries are required to abolish entry visas and permits for ECOWAS nationals, adopt the ECOWAS travel certificate and implement the ECOWAS Brown Card scheme.
- **Free Movement of Goods**: In line with the harmonization of customs documents by ECOWAS Member States, member countries have to adopt the ECOWAS certificate of origin, customs nomenclature (HS) and the customs declaration form. Member countries are expected to be charging the Community levy of 0.5% on imported goods from countries outside the region. Member countries are also to fully implement the ECOWAS Trade Liberalization Scheme (ETLS) under which unprocessed goods from the region enter member countries duty-free. However, industrial products must satisfy ECOWAS rules of origin and other certification conditions before being eligible for ETLS benefits.
- **Common External Tariff**: As a step towards the conclusion of an ECOWAS customs union, and in pursuit of further regional integration, the ECOWAS countries are committed to the establishment of a Common External Tariff (CET). This aspires to harmonize the tariff structures of the Anglophone countries with that of the common external tariff currently upheld by the francophone UEMOA countries.

1.5.3 Economic Partnership Agreement (EPA) with the EU

Until 31 December 2007, about 97% of the exports of the ACP countries entered the European Union (EU) duty-free in accordance with the Cotonou preferential trade regime. The WTO waiver, under which the Cotonou regime was permitted, expired on 31 December 2007. To retain such duty-free access beyond this point African countries, with particular reference to West African (WA) countries (ECOWAS + Mauritania), had to negotiate a WTO-compatible Economic Partnership Agreement (EPA) with the EU. Under the new regime, the WA region and the other five ACP regions negotiating EPAs with the EU are expected to also open their markets progressively in an asymmetrical manner to EU exports. This will be accompanied by EU-supported measures to accelerate West Africa's integration processes, upgrade its industries and build its productive capacity and competitiveness, which have implications for commodity value chain development. This is to enable West Africa to take full advantage of market access opportunities offered by the EU and to integrate into the global economy.

1.5.4 Bilateral Trade Relations and Commodity Value Chain Development

Bilateral agreements with the objective of taking advantage of opportunities offered in key strategic markets, fostering business cooperation and technology transfer and encouraging Foreign Direct Investment (FDI) are critical for supply chain management in Africa and commodity value chain development. On Preferential Trade Arrangements, Ghana benefits from the Generalized System of Preferences (GSP) offered by industrialized countries other than the EU and the USA. The USA offers a number of African countries special access to its market under the African Growth and Opportunity Act (AGOA). Under AGOA over six thousand products from the beneficiary countries enter the U.S. market duty-free and quota-free. Apparel made from U.S. fabric, yarn or thread also enters the U.S. market duty-free and quota-free. This means that commodity value chains go beyond national boundaries.

1.6 Food Safety

There are food safety concerns with vegetables generally. Issues of salmonella and *Ecoli* associated with vegetables have become a primary concern in the food industry. This greatly affects the tomato industry, as people become less willing to patronize tomatoes for salad and so on. The increasing use of agrochemicals by farmers in the production of vegetables also has serious food safety concerns. Most vegetable farmers either use the wrong chemicals at the wrong times or overuse the right chemicals all of which make vegetables unwholesome for consumption. Instances of harvesting of fresh vegetables for the market soon after chemical application are rife. Most farmers also produce under poor sanitary farm conditions. Post-harvest handling of vegetables and tomatoes for that matter is also done under poor sanitary and phytosanitary conditions in most cases raising food safety concerns among consumers, especially those in the high-value market segment. Critical efforts are needed to promote food safety standards across the different segments of the commodity value chains in Africa.

1.7 Technology and Innovation

The use of production technologies such as greenhouses is very limited in Africa and this affects supply and commodity value chains negatively. Farmers still engage in traditional methods of production using low yielding varies and rudimentary farm practices. For example, the Enviro Dome Greenhouse System introduced in Ghana by the Forest and Horticultural Crops Research Centre (FOHCREC) of the University of Ghana in collaboration with Stevicksen Ventures Ltd (SVL) and Eisenberg-Agri (Beijing) Ltd as a catalyst for all year-round production of vegetables such as tomatoes is yet to gain widespread use by farmers.

1.8 Market and Governance Structure

The market and governance system in most commodity value chains in Africa is not well structured. At the farm level, there are no strong producer collectives with well-defined and functioning governance systems that regulate the activities of producers. The lack of properly constituted governance system in the industry makes it difficult for farmers to produce to meet set standards required by high-value end markets. Thus every producer does what he or she wants and this affects the overall performance of the production segment of commodity value chains. Marketing is based on bargaining with no formally defined rules and regulations or governance structure. Farmers harvest and sell as and when the product is ready for the market and buyers are available. The big supply and consumer markets are however, managed and controlled by 'market queens' who have informal governance structure for market operations. Thus producers are unable to directly sell their produce in such markets without passing it through the 'market queens'. Again, the forces of supply and demand largely determine prices with the 'market queens' sometimes manipulating the process through collusion. For example, the 'queens' realizing that farmers entered into contract with processors or high-value end market actors can intentionally offer higher prices within a very short period just to entice farmers to breach such contracts. Once farmers breach the contracts, the 'queens' then drop the prices below the contract prices all in attempt to short-chained the farmers. Farmers being driven by short-term revenue maximization motivations fall victims to this most of the time. This will only change with sensitization of farmers on the need to operate within robust FBOs and to adhere to signed contracts for long-term benefits.

References

Akudugu, A. M. (2011). Rural banks' financial capital and livelihoods development of women farmers in Ghana. *Journal of Enterprising Communities: People and Places in the Global Economy, 5*(4), 248–264.

Banchirigah, S. M. (2006). How have reforms fuelled the expansion of artisanal mining? Evidence from Sub-Saharan Africa. *Resources Policy, 31*(3), 165–171.

Banchirigah, S. M. (2008). Challenges with eradicating illegal mining in Ghana: A perspective from the grassroots. *Resources Policy, 33*(1), 29–38.

Block, J. R. (1994). A new view of agricultural productivity in Sub-Saharan Africa. *American Journal of Agricultural Economics, 76*(4), 619–624.

Burkhard, B., & Maes, J. (2017). *Mapping ecosystem services* (p. 374). Sofia: Pensoft Publishers.

Costanza, R., de Groot, R., Braat, L., Kubiszeewski, I., Fioramonti, L., Sutton, P., ... Grasso, M. (2017). Twenty years of ecosystem services: How far have we come and how far do we still need to go? *Ecosystem Services, 28,* 1–16.

Cunningham, G. (2012). *Community economic lietracy and the "Leaky Bucket"*. Coady International Institute, St Francis Xavier University.

Dewbre, J., & Borot de Battisti, A. (2008). *Agricultural progress in Cameroon, Ghana and Mali: Why it happened and how to sustain it* (OECD Food, Agriculture and Fisheries Working Papers 9), 60.

Diao, X., Hazell, P., Resnick, D., & Thurlow, J. (2007). *The role of agriculture in development: Implications for Sub-Saharn Africa.* Retrieved from Washington, DC.

Díaz, S., Pascual, U., Stenseke, M., Martín-López, B., Watson, R. T., Molnár, Z., ... Shirayama, Y., (2018). Assessing nature's contributions to people. *Science, 359*(6373), 270–272.

Garcia-Winder, M., Riveros, H., Pavez, I., Rodriguez, D., Lam, L., Arias, J., & Herrera, D. (2009). Agrifood chains: A tool for strengthening the institutional framework of the agricultural and rural sector. *Perspectives,* 26–39.

Fan, S., Omilola, B., & Lambert, M. (2009). *Public spending for agriculture in Africa: Trends and composition* (ReSAKSS Working Paper, 28), 20.

FAO. (2000). *The state of food and agriculture: Lessons from the past 50 years.* Rome, Italy.

Frisvold, G., & Ingram, K. (1995). Sources of agricultural productivity growth and stagnation in Sub-Saharan Africa. *Agricultural Economics, 13,* 51–61.

Fulbright, N. R. (2019). *Climate change and the food and Agribusiness sectors: Adaptation, mitigation and innovation.*

Fulginiti, L. E., Perrin, R. K., & Yu, B. (2004). Institutions and agricultural productivity in Sub-Saharan Africa. *Agricultural Economics, 31*(2–3), 169–180.

Hilson, G. (2002). The environmental impact of small-scale gold mining in Ghana: Identifying problems and possible solution. *The Geographical Journal, 168*(1), 57–72.

Huge, J., Rochette, A. J., Bethune, S., Parra Paiton, C. C., Vanderhaegen, K., Vandervelden, T.,... Janssens de Bi Bisthoven, L. (2020). *Ecosystem Services,* 42 (2020) 101079.

IPBES. (2019). *Assessment report on pollinators, pollination and food production.* https://ipbs.net/assessment-reports/pollinator.

InterAcademy Council. (2004). *Realizing the promise and potential of African agriculture: Science and technology strategies for improving food security and agricultural productivity in Africa.* Amsterdam, The Netherlands: Royal Netherlands Academy of Arts and Sciences.

Kuah, A. T. H. (2002). Cluster theory and practice: Advantages for the small business locating in a vibrant cluster. *Marketing and Entrepreneurship, 4,* 206.

Lusigi, A., & Thirtle, C. (1997). Total factor productivity and the effects of R&D in African agriculture. *Journal International Development, 9,* 529–538.

Porter, M. (1998). Clusters and the new economics of competition. *Harvard Business Review, 76,* 77–90.

Porter, M. (2008). *Clusters, innovation, and competitiveness: New findings and implications for policy*, Stockholm.

Rao, D. S. P., & Coelli, T. J. (1998). *Catch-up and convergence in global agricultural productivity 1980–1995.* Department of Econometrics, University of New England, Armidale.

Suhariyanto, K., Lusigi, A., & Thirtle, C. (2001). *Productivity growth and convergence in Asian and African agriculture.* New York: Palgrave.

Tasser, E., Schirpke, U., Zoderer, B. M., & Tappeiner, U., (2020). Towards an integrative assessment of land-use type values from perspective of ecosystem services. *Ecosystem Services*, 42 (2020), 101082.

Thirtle, C., Hadley, D., & Townsend, R. (1995). Policy-induced innovation in Sub-Saharan African agriculture: A multilateral Malmquist productivity index approach. *Development Policy Review, 13*, 323–348.

Webber, C. M., & Labaste, P. (2010). *Building competitiveness in Africa's agriculture. A Guide to Value chain Concepts and Applications*. The World Bank, Washington, DC.

World Economic Forum. (2019). *System initiative on shaping the future of food. Innovation with a purpose: Improving traceability in Food value chains through technology innovation.*

CHAPTER 2

Commodity Value Chain Structures

Abstract Understanding the key structures of commodity supply and value chains is critical in ensuring them (chains) robust. Whiles the development of some supply and value chains should only be limited to strengthening local or cluster structures, others have to go beyond that to strengthen their competitive positions in national and international structures. So this chapter focuses on examining the different commodity value chain structures that affect performance. The chapter argues that value chain actors must carefully consider their competitiveness in relation to segments and types of markets—domestic or international to focus on. They must understand what goes on in each segment of the chain or market to be able to competitively take advantage of the opportunities therein.

Keywords Actors · Commodity · Consumers · Value chains · Market channels

2.1 Domestic Market

Naturally, the key end markets for most agricultural commodities in Africa are the domestic ones where the primary market segments are the open-air retail market and a much smaller, but growing, supermarket channel.

A.-R. Alhassan and M. Abunga Akudugu,
Supply Chain Management in African Agriculture,
https://doi.org/10.1007/978-3-030-54209-2_2

Given Africa's history of agricultural production and existing domestic market demands, opportunities for import substitution exist when local growers can offer an improved quality and extend their growing season. Currently, food imports dictate the price trend for local food produce in most of Africa. Supermarket's channels are dominated by imports due to their quality and shelf life and thus imported food and food products set a bar for the local ones in this market segment.

It is anticipated, as it is across the world that foods such as fresh vegetables consumption will continue to increase as the culture of consuming fresh vegetables as a healthy food takes hold and as the health benefits of these become more recognized. This might result in a visible increase in the demand for fresh foods in local as well as regional markets requiring better supply value chain management. In addition, as incomes increase, a growing number of middle-class Africans are developing taste for fresh rather than processed foods.

2.2 Competitive Position in Domestic Markets

Presently, imported foods and food products constitute the main competition in the domestic markets in Africa. For example, Ghana spent approximately US$100.3 million importing tomato paste, fresh and peeled tomatoes in 2011, out of which $92.1 million was spent on tomato paste. The country has a competitive advantage in the fresh tomato domestic market as much of the fresh tomatoes consumed in Ghana are locally produced. The supply however, varies across the different agroecological zones with some overlaps. It is only from sometime in December to May that imports of fresh tomatoes from Burkina Faso supplement local supply due to unfavourable weather conditions for tomato production in the country around that time. Unfortunately, the use of climate smart agricultural production technologies such as greenhouses is limited, as only a few farmers employ these technologies. During the local peak production seasons imported foods such as tomatoes are not present in the market. Thus direct competition between local and imported foods tomatoes typically takes place during the period of December to May when tomatoes are imported to substitute for the inadequacy of local supply at this time. Although consumers prefer the taste of local tomatoes, Ghanaian tomatoes cannot compete on quality with most imported tomatoes as the imported ones appear to have better visual appeal due to grading. The imported fresh tomatoes are also said to have a longer shelf life due to use of pre-cooling and being treated with calcium or other firmness enhancers, in addition to the individual varietal qualities.

Local tomatoes are more competitive than imported tomatoes in the open-air retail markets where quality requirements such as longer shelf life and visual appearance are lower and sanitary and phytosanitary standards are not enforced. However, the inability to meet quality requirements of local supermarkets, limits local producers competitiveness vis-a-vis imports in this high-value end market channel.

2.3 Consumer Preferences

During the off-season domestic consumer demand is not very sophisticated, with consumers generally content with what is available. The price difference in the beginning of season relates to enthusiasm for the reappearance of local food crops and the fact that local buyers believe that the imported foods have no taste. Demand for specialty varieties of food crops such as tomatoes, is currently limited to the supermarket's sector. There is currently potential for introducing several new varieties for consumer preference testing. The strategy of introducing new higher value crop varieties onto the shelves should be guided by consumer preference studies in partnership with supermarkets, paired with local farmers introducing changes into their production technology. Some consumers prefer open-field crops to greenhouse crops, which in their opinion do not seem to have the natural taste. Besides, consumers are of the opinion that open-field crops are healthier, containing less pesticides and fertilizers. Also some consumers think that the open-field crops are much richer and more characteristic to what they have been used to through the years. Other comparative characteristics of local and imported foods, particularly tomatoes are shown in the Table 2.1.

2.4 Market Opportunities

Opportunities exist for import substitution in the domestic market focusing on the open-air retail market and the supermarket segments as well as on-farm and off-farm processors. This can be accomplished by extending the production season for local tomatoes. With slight and not very costly production improvements such as effective scheduling of planting time, introduction of efficient production technologies and choosing the right varieties, the trading season for local tomatoes can

Table 2.1 Consumer preferences under section 2.3

Price	*Taste*	*Shape*	*Colour*	*Consistency*
In most cases the prices of imported tomatoes are lower than locally produced tomatoes	The view of the Ghanaian consumer is that imported tomatoes 'have no taste' which is one of the key competitive advantages of local tomatoes	Imported tomatoes are mostly uniform in shape compared to local tomatoes	Imported tomatoes have a uniform pale red colour compared to local tomatoes that mostly have a non-uniform colour	The common characteristics for all of the imported varieties are greater firmness, compared to local tomatoes

be extended to both dry and rainy seasons. In this way, local producers will gain extra months of product marketing. In the domestic open-air market segment the competition is currently based on price. Quality requirements are not stringent and can usually be met by local producers. There are no specific requirements for shape, product uniformity, visual appeal and shelf life. Transactions in this market segment are informal with no invoices used. Phytosanitary requirements are poorly enforced at the open-air market level. In most open-air markets there are no sanitary inspection authorities for inspection of tomatoes and agricultural commodities generally before the products reach the retail markets. There are no proper traceability mechanisms, as local authorities do not certify farmers and farms but this has to change as efforts are being made to improve the performance and productivity of the entire tomato value chain. This is particularly important for those selling directly to supermarkets, as they must produce to meet specific requirements before their produce is accepted.

Since the entry requirements in the open-air market are not stringent and local consumer taste preferences already favour local tomatoes over imported, there is an immediate opportunity to substitute for imports by increasing the local production season through improvements in production systems including adoption of greenhouse technology, irrigation facilities (drip irrigation), minimizing production costs and increasing yields.

The domestic supermarket channel is where competition with imports is strongest and quality requirements are strict. That makes competing in this market channel a longer-term goal for local producers. Tackling that will require improved quality and sufficient volumes supplied on a sustained basis, which will require pre-cooling/cold storage, grading and stronger cooperation among producers. Farmers need to respond to the requirements of supermarket suppliers for on time delivery, formal invoices, consistency of supply and uniformity and longer shelf life of the product. Specific requirements of domestic supermarkets include:

1. Longer shelf life (at least 1 week)
2. Seasonal contracts and ability to deliver on time and to volumes contracted on a consistent basis
3. Invoiced transactions
4. Ability to accept post-payment terms
5. Quality certifications and compliance with the required standards
6. Single point of transaction (which requires farmer organization and strong relationships between growers and distributors /supermarket suppliers).

2.5 Competitive Position in Exports Markets

Exports of foods and food products including fresh vegetables such as tomatoes are driven by demands of supermarkets in destination countries. Africa's presence in the higher value and rapidly growing supermarket segment is currently almost non-existent due to inability to meet quality and volume requirements of supermarket buyers. Currently, Africa competes on price due to the preferential tariff regime for its exports to EU countries and the USA, as compared to other exporters. In terms of tomatoes, Africa is currently not occupying a competitive position in the export market as it exports very little of what is produced to neighbouring countries within the continent. Most consumers in the destination countries prefer the taste of African vegetables and other agricultural produce. This is however, likely to change as consumers get used to higher quality products from other countries. Without tariff advantages African producers would need to drastically improve the quality of product, taste and consistency of supply volumes in order to be competitive in export markets.

2.6 Market Channels

African foods and food products reach the final consumer through several channels. This is defined by the way the retail distribution channels in end markets are currently structured. There are five main market channels operating across Africa and these are briefly discussed below.

2.6.1 Open-Air and Roadside Retail Markets

This market channel currently dominates the retail channel for fresh foods and even processed food products in terms of sales volumes. It is estimated that about 80% of fresh foods and food products produced in the continent pass through this channel. Open-air markets handle large volumes of produce, hence competition is highly price related and sale prices are low compared to other channels. Quality requirements in this channel are not sophisticated and competition is mainly based on price. Consumers in this segment choose the product that they feel offer the best value proposition (price in relation to appearance). Transactions in this channel among producers, traders and consumers are mostly informal, with no invoices or receipts used. A significant issue also is the lack of enforcement of sanitary and phytosanitary standards. Consumers prefer to purchase foods in this channel due to an overall perception that the produce here is better quality and fresher than in supermarkets where produce stays on the shelf for a longer time. At the height of the production season temporary wholesale open-air markets are established in production centres where local producers sell their produce to traders. At the local open-air markets, growers sell their produce to aggregators/wholesalers to the major market centres.

2.6.2 Supermarkets

Major supermarket chains including Shoprite, Game and others dominate this channel. Although fruit and vegetables occupy small vending spaces within typical African supermarkets, the supermarket channel has been growing in recent times as food sales outlet and is projected to continue to grow. Supermarkets sell both local and imported fresh foods and processed food products. It must be noted that this channel is the most regulated one in terms of quality, sanitary standards and supplier requirements. So the capacity of local producers needs to be built to be able to meet the volume requirements, delivery schedules and product standards established by the supermarkets. Farmers need to improve their product quality and consistency of production volumes and establish relationships with distributors and traders that supply supermarkets. This is critical because supermarkets are looking for consistent supply from a limited number of trusted suppliers and also this is the channel where competition from imports is the strongest.

2.6.3 Small Grocery Stores

This channel includes small local grocery/convenience stores and kiosks that trade smaller volumes and offer less variety at prices marginally higher than the open-air retail markets. This channel serves consumers who are looking to purchase a mix of different food products and beverages in one nearby location, in addition to fresh produce. Many of these stores do not offer a full range of fruits and vegetables, which occupy small proportions of their total shelf space.

2.6.4 Processing Companies

Processing companies are key actors of commodity value chains. Processors are expected to collect fresh produce from farmers and/or aggregators for processing with production structured in such a way that processing companies do not experience difficulty getting sufficient supply from growers on a consistent basis. To guarantee constant supply, processors need to sign forward contracts with farmers and farmer-based organizations (FBOs). To stop side selling by farmers, a basic amount has to be agreed upon before harvest and if prices at harvest are higher than agreed, farmers will have the opportunity to renegotiate on win–win basis.

2.6.5 Institutional Markets

Producers, processors and traders supply commodities such as cereals, legumes, roots and tubers as well as vegetables to institutional buyers including schools, hospitals, restaurants and hotels, among others. Consumers in this market segment demand high standards and consistent supplies. Thus the capacities of producers, traders and local processors need to be built so that they could produce to meet the sanitary and phytosanitary standards as a primary requirement in this market segment. Besides, producers need to be helped to establish relationships with buyers in this market so that they will produce to meet their requirements.

2.7 Actors and Functions

Different actors with varying degrees of influence perform the different functions along commodity value chains. These actors are described in the following subsections.

2.7.1 *Inputs and Services Suppliers*

Inputs and services suppliers play important functions in the overall performance of supply chains and commodity value chains through the supply of inputs and services necessary for production. They are critical in ensuring farmers have access to all the necessary inputs for tomato production. They sometimes supply inputs on credit to farmers to produce and pay upon harvest. Some input dealers also provide capacity building services on good agricultural practices and better ways of using inputs they supply. Their influence on the overall performance of supply and commodity value chains is high.

2.7.2 *Producers*

The majority of producers/farmers across Africa are smallholders with individual production areas of less than a hectare. The total number of smallholder producers in Ghana for example is well over 1 million managing thousands of hectares of land put under crop production. Data on the total number and size characteristics of farmers are however, not readily available. Harvesting of crops in case of smallholder farmers is done directly by producers without hiring any outside labour. The producers mostly do basic sorting and grading manually before selling to buyers. The post-harvest handling nodes of commodity value chains are missing a warehousing function with the ability to extend shelf life. Pre-cooling/cold storage/packing facilities for vegetables such as tomatoes are lacking in production centres. A cold storage facility (including a forced-air cooling system) would extend the shelf life of produce, and allow for storing during the peak season when price is low, and provide the opportunity to meet an extended supply calendar required by consumers. Red tomatoes for instance can be stored for 4–5 weeks in cold storage allowing for extended supply of tomatoes thereby substituting for imports. Furthermore, proper packing, grading and sorting at such a facility will allow boxing of standard quality and uniform products. The best quality boxes can go to local supermarkets and the low-quality ones to the other market segments such as open-air retail markets where quality requirements are lower. Truck sourcing at the packing and cooling facility will be more efficient than the current procedure of going to production villages for days and negotiating with many different farmers for commodities. Women provide a lot of the labour force involved in all the stages of crop production across Africa—from planting to manual harvesting, grading and marketing. The level of influence of producers on the performance of the overall value chains is limited because they have little control over resources, prices and markets.

2.7.3 Intermediaries

Before reaching the final consumer, agricultural commodities go through a set of intermediaries who act as aggregators or transporters to end markets. Small producers frequently sell their produce at the farm gate to traders who transport it by their own means and sell on the wholesale or retail markets.

a. Lead boys

Lead boys play critical roles in performing key functions in agricultural commodity value chains as they link up producers to buyers. Their influence on the performance of the value chains is high because they determine whom the traders should buy from. They also play a key role in price negotiations and facilitate the movement of produce from one area to another as they mostly act as translators for the buyers and the farmers.

b. Market queens

The largest volume of domestic commodity sales goes through market queens, who operate only on the local markets supplying mainly the open-air market channel, but also some supermarkets and small grocery stores.

In Ghana alone for example, it is estimated that there are over 1000 market queens across different markets in the country. Market queens make sporadic transactions when they see a market opportunity, and rarely have long-term relationships with particular farmers. The market queens usually have a maximum of 3 days to sell the products. On the fourth day of trading the risks are very high that the produce will spoil. Market queens have strong influence on the performance of the commodity value chains as they control entry and distribution of produce in markets under their jurisdictions.

c. Distributors

Distributors usually buy agricultural commodities from producers and supply supermarkets and small grocery stores. These distributors are performing sorting, grading and packaging of comodities to meet supermarket requirements and usually make a margin per kilo of packaging. To meet supermarket requirements for steady supply, distributors act as consolidators of produce from a large number of small producers. They have high influence on the commodity value chains particularly in the high-value end markets.

d. Transporters

Transporters play important roles in agricultural commodity supply management and value chains as they facilitate the movement of inputs to producers and outputs to consumers. Thus they ensure that produce from the farmer reaches the final consumer. They have high influence on the performance of commodity value chains.

e. Retailers

Retailers perform critical roles as they serve as the link between wholesalers/farmers and consumers. They however, have limited influence on the performance of commodity supply and value chains as they possess little control over resources and rely on traders for supply, sometimes on credit.

f. Importers

Companies owned completely by or in partnership with local businessmen import fresh and processed agricultural produce into African countries. The number of such importing companies in any particular country is usually not well known. Some importers sell to distributors who then supply supermarkets directly and other market outlets. Some supermarket chains also run their own import activities for fruits and vegetables.

g. Processors

Although there may be large processing companies in Africa, most of which were set up at decades ago, most of them are currently out of operations. Packaging companies that import food products and package them in country dominate the processing industry. Factories put up to process tomatoes and meat in Ghana for example are not operating because they are unable to get sufficient supply from producers on consistent basis. Small-scale on-farm processing companies exist and are currently operating although on a very small scale. Such small-scale processing companies try to develop contract-based relationships with farmers, but even in cases when a contract is signed, processors report that farmers often choose to side-sell if the market price is more favourable. The level of influence of processors on the performance of agricultural commodity supply and value chains is currently not high.

h. Consumers

Consumers are expected to play important roles in the overall performance of agricultural commodity supply and value chains as they dictate what should be produced, how and when it should be produced. Unfortunately, the preferences of consumers are not always transmitted back along the supply and value chains to influence the type of commodities grown and production practices adopted by farmers.

2.8 Overall Status of Commodity Value Chains in Africa

The current supply and commodity value chains are not robust, as the various segments of the chains are not functioning effectively. There is no equity in terms of distribution of profits across the different segments of the chains. The different actors along the chains make different levels of profits due to the different levels of risks. Thus the productivity of the chains is low and this current status is seen in the SWOT analysis Table 2.2.

Table 2.2 SWOT table on overall status of commodity value chains in Africa

Strengths	*Weaknesses*
1. Large population and high per capita consumption of food products	1. Most consumers are price sensitive and ignorant of quality requirements
2. A high ratio of agricultural land to agricultural population	2. Significant cost variation in production across clusters, sizes and types of farms
3. Agribusinesses and service providers for agricultural production already exist	3. Poor quality of commodities produced by farmers
4. Willingness of farmers to adopt new production technologies that will lead to increased productivity and income levels	4. Traders operating without regulations leading to significant value destruction
5. Traders are willing to invest in improving the performance of commodity value chains given the right support	5. Low agricultural productivity and little productivity support services to farmers
6. There is a growing middle class of consumers and this is good for the fresh food industry as demand is set to increase	6. Farmers not organized with general lack of proper management structures
	7. Poor quality extension and breeding services
	8. Support structures and industry are not well developed to support agricultural productivity
	9. Seasonality leading to fluctuations and uncertainty in commodity supplies and quality
	10. Underdeveloped cold chains leading to high losses of fresh produce in transit
	11. Poor physical infrastructure such as impassible roads linking production and consumption centres

(continued)

Table 2.2 (continued)

Strengths	*Weaknesses*
	12. Small-scale and fragmented farm holdings
	13. Lack of remunerative producer prices for agricultural commodities
	14. Low utilization of installed capacity of processing plants
	15. Lack of well-defined national and continental policies for agricultural commodity supplies value chain development generally
Opportunities	*Threats*
1. Consumption can be increased with the right emphasis on nutritional value, quality and price. Capitalize on nutritional value placed on consumption of fresh organic farm produce and related products by middle class	1. Farmer cooperation around collection and central production points could be expensive and difficult to manage
2. Potential to improve yields through better production technologies availability and breeding	2. Lack of varietal management at the macro level and low potential of indigenous crop varieties and animal breeds
3. Potential to maintain/establish feasible, larger farms and interest in investing in larger model farms	3. Unregulated imports of agricultural produce and related commodities
4. Induce productivity and quality-specific farmer cooperatives around production and possibly aggregation, and support efficient collection at larger scales	4. Vested interests in perpetuating the dependence on imports of agricultural produce and related commodities
5. Willingness of all farm types to invest in productivity	
6. Large aggregate production of commodities	
7. Availability of rural and urban markets to sell farm produce	
8. Joint ventures based on large potential and access to regional markets	
9. Government's interest and willingness to contribute	
10. There exist institutions that could develop and implement quality standards	

CHAPTER 3

Commodity Clusters, Arenas, Linkages and Business Models

Abstract Transformation of African agriculture requires strategic interventions that are economically viable, financially feasible, environmentally sustainable, socially acceptable, culturally agreeable, gender sensitive, generationally stable and politically neutral. Such strategic interventions must have at the core of it the pooling of resources for coordinated action around the centre—clustering. The clusters will allow proper functioning of the various actor arenas with proper linkages to innovative business models for success. This chapter thus presents some useful suggestions on how clusters and products or commodities are identified for strategic interventions in the various arenas for value addition through vertical and horizontal linkages that promote chain and actor equity for sustainable and inclusive development.

Keywords Agricultural commodities · Arenas · Business models · Clusters · Value addition

A.-R. Alhassan and M. Abunga Akudugu,
Supply Chain Management in African Agriculture,
https://doi.org/10.1007/978-3-030-54209-2_3

3.1 Identification of Commodity Clusters for Intervention

Farming takes place in most parts of Africa but there are some key production areas for specific agricultural commodities in country and across countries. The identification of clusters must be done consistent with these key production areas using a multistage process based on both current available production data and areas with production potentials. At the continental level, the map of clusters is divided into five categories in accordance with the four subregional groupings—West Africa, East Africa, Central Africa and Southern Africa.

The clusters are made up of countries within each of the subregional blocs that must work together to develop both in-country and cross-country commodity supply and value chains. Farmers in each of these clusters engage in both rain-fed and irrigated production. Those into rain-fed production only operate during the rainy season with the irrigated ones operating in the dry season using water from dams and rivers. This cluster has potentially millions of farmers who are into crop farming, animal rearing and aquaculture. However, most farmers are currently not doing well due to poor performance of the agricultural sector. Majority of farmers are smallholder, cultivating an average of 0.5–2ha with an estimated harvest of about 6t/ha. Given the necessary productivity improvements support and market linkages, farmers will be able to produce more than twice their current output levels.

The key country and continental level service providers in these clusters must be identified for provision of services for farmers and other actors along the value chains for improved performance. For instance, country level agricultural ministries must provide irrigation, mechanization and extension services to farmers. These ministries must enter into private–public-partnerships in the provisioning of some of these services. For example, private individual tractor operators could work with these ministries to provide land preparation services to farmers based on win–win arrangements for all parties involved. The other service providers are the financial institutions for the provision of financial services. Financial and non-financial Non-Governmental Organizations (NGOs) operating in the clusters also provide some critical services including extension and credit to actors along commodity supply and value chains. There are also private transport owners for provision of transport services to farmers and traders as well as auxiliary service providers such as carpenters for carpentry services.

The input dealers at country, cluster and continental levels must also be positioned to provide inputs services to farmers and actors along the chains. The key traders of farm produce across Africa are the market queens, open-air market retailers, roadside retailers and grocery stores, particularly for processed foods. These actors must have their capacities enhanced for improved performance of commodity supply and value chains. Academic and research institutions must also be involved in the generation of critical knowledge, improved production technologies and farm practices for increased productivity.

3.2 Cluster Identification for Tomato Production—Ghana Case Study

Tomato is produced in most parts of Ghana but there are some key production areas. The identification of clusters involved a multistage process based on both current available production data from MOFA and areas with production potentials. For the purposes of this report, the map of tomato clusters in Ghana is divided into five categories in accordance with the agroecological zones of the country as outlined in the following subsections.

3.2.1 Navrongo Area Tomato Cluster

The cluster is made up of farmers in the Kasena-Nankana Municipality, Kasena-Nankana West, Builsa South and North Districts. Farmers in this cluster engage in both rain-fed and irrigated tomato production. Those into rain-fed tomato production only operate during the rainy season with the irrigated ones operating in the dry season using water from dams (e.g. Tono Irrigation Dam) and rivers. This cluster has potentially more than 5000 farmers who are into vegetable farming including tomatoes. However, most farmers are currently not into tomato production due to poor performance of the sector. The farmers cultivate an average of 0.5 ha with an estimated harvest of about 6t/ha. Given the necessary productivity improvements support and market linkages, farmers will be able to produce more than twice their current output levels.

The key service providers in this cluster are ICOUR for provision of irrigation, MOFA for the provision of extension services and private individual tractor operators for land preparation services. The other service providers are the Naara Rural Bank Ltd, Agricultural Development Bank and GCB Bank Ltd for the provision of financial services. Financial and non-financial Non-governmental Organizations (NGOs) operating in the cluster also provide some critical services including extension and credit to actors along the tomato value chain. There are also private transport owners for provision of transport services to farmers and traders as well as carpenters for carpentry services.

The input dealers are Sule Issah Agrochemical shop, Jagula Agrochemical Enterprise, Amidu agrochemical shop and Dapat Royal Agrochemicals shop. These input dealers get their supplies from Simple Prince agrochemicals (Bolga), Meridian seeds (Bolga), Ganorma and Wumpini agro-inputs (Tamale) and Sefa and Jane (Kumasi). The key traders are the market queens, open-air market retailers, roadside retailers and grocery stores, particularly for processed tomatoes. Northern Star Tomato Company although currently not operating could be a driving force of the tomato value chain in this cluster due to its proximity.

The Navrongo Health Research Centre (NHRC), the Savannah Agricultural Research Institute (SARI), the Damongo Agricultural College (DAC) and University for Development Studies (UDS) are also key players. The NHRC conducts research on health, particularly farm and water-related diseases that affect the performance of actors along the tomato value chain in the cluster. SARI although outside the cluster plays critical roles in breeding and releasing quality tomato varieties/seeds that are well-adapted to the climatic and environmental conditions of this cluster. The UDS particularly the Faculty of Applied Sciences, Faculty of Agribusiness and Resource Economics and Faculty of Agriculture conduct critical research for value chain enhancement. DAC (outside this cluster) is also a key actor in the training of technical officers specialized in vegetable production using modern agricultural technologies such as greenhouses and irrigation. Bolgatanga Polytechnic (B-Poly) is also a key play in training of people specialized in logistics and transport.

3.2.2 Bolgatanga Area Tomato Cluster

This cluster is made up of farmers from Bolgatanga Municipality, Bongo, Tanlensi and Nabdam districts. Farmers in this cluster engage in both rain-fed and irrigated vegetable production notable among which are tomatoes, leafy vegetables, pepper, okra and onions. Some of the farmers are into both rain-fed and irrigated vegetable production. Those into rain-fed tomato production only operate during the rainy season with the irrigated ones operating in the dry season using water from dams (e.g. Vea Irrigation Dam) and the White Volta and its tributaries. This cluster has potentially more than 6000 tomato farmers most of who are moving away from tomato production due to crop failures as a result of pests and diseases as well as lack of proper marketing mechanisms. Most farmers cultivate less than a hectare with an estimated harvest of about 5tons/ha annually. Given the necessary productivity improvements support and market linkages, farmers will be able to produce up to three times their current output levels.

The key service providers in this cluster are ICOUR for provision of irrigation, MOFA for the provision of extension services and the Agricultural Mechanization Services Centre (AMSEC) as well as private individual tractor operators for land preparation services. There is an array of financial actors in this cluster that provide financial services for the financing of activities across the tomato value chain. These actors include the Builsa Community Rural Bank Ltd, Naara Rural Bank Ltd, Bongo Rural Bank Ltd, Agricultural Development Bank, National Investment Bank and GCB Bank Ltd as well as non-bank financial institutions such as Bayport Financial services and Sinapi Aba for the provision of financial services and private transport owners for provision of transport services to farmers and traders as well as carpenters for carpentry services. Financial and non-financial NGOs operating in the cluster also provide some critical services including extension and credit to actors along the tomato value chain. The input dealers are Simple Prince agrochemicals and Meridian seeds, among others. Northern Star Tomato Company although currently not operating could be a driving force of the tomato value chain in this cluster. Although located outside this cluster, NHRC, SARI, UDS, DAC and B-Poly are also key players. Bolgatanga Technical University is also a key player in training of people specialized in logistics and transport.

3.2.3 Bawku Area Tomato Cluster

This cluster is made up of farmers in the Bawku Municipality, Bawku West, Binduri, Pusiga and Garu-Tempane districts. Farmers in this cluster engage in both rain-fed and irrigated vegetable production and are known for onion production but many of them also produce tomatoes, pepper, okra and leafy vegetables. Most of the farmers engage in both irrigated and rain-fed tomato production. They irrigate using water from small-scale irrigation dams/dugouts and rivers, notably the White Volta Basin and its tributaries. This cluster has potentially more than 10,000 tomato farmers most of who are currently not into tomato production due to poor performance of the sector. The farmers cultivate an average of 0.5 ha with an estimated harvest of about 6t/ha annually. Given the necessary productivity improvements support and market linkages, farmers will be able to produce more than twice their current output levels.

The key service providers in this cluster are MOFA for the provision of extension services and private individual tractor operators for land preparation services. The financial actors in this cluster that provide financial services for the financing of activities across the tomato value chain are the GCB Ltd, ADB, BESSFA Rural Bank Ltd, Toende Rural Bank Ltd, Bayport Financial Services, etc. There also exist private transport owners for provision of transport services to farmers and traders as well as carpenters for manufacture and maintenance of boxes for carrying tomatoes. Financial and non-financial NGOs operating in the cluster also provide some critical services including extension and credit to actors along the tomato value chain. Many private input dealers supply fertilizers and agrochemicals to farmers. Northern Star Tomato Company although currently not operating could play critical roles in improving the performance of the tomato industry in this cluster because of proximity. Although located outside this cluster, NHRC, SARI, UDS, DAC and Bolgatanga Technical University are also key players. Note that SARI has an outstation in this cluster, located at Manga in the Bawku Municipality.

3.2.4 Tamale Area Tomato Cluster

This cluster is made up of farmers from Tamale Metropolis, Savelugu-Nanton, Kumbungu, Tolon and Sagnerigu districts. Farmers in this cluster engage in both rain-fed and irrigated vegetable production. Most of the farmers engage in rain-fed tomato production as a result of the fact that farmers in the region are not so much into irrigation except those around the Botanga and Libga areas, among others. This cluster has potentially more than 15,000 vegetable farmers including tomatoes with most farmers into pepper and leafy vegetable production. The farmers cultivate an average of 0.5 ha with an estimated harvest of about 4tons/ha annually. Given the necessary productivity improvements support and market linkages, farmers will be able to produce more than three times their current output levels. The key service providers in this cluster are the Ghana Irrigation Development Authority (GIDA) for the provision of irrigation, MOFA for the provision of extension services and private individual tractor operators for land preparation services.

There is a vast array of financial services providers in this cluster including the GCB Ltd, ADB, Barclays/Absa Bank, NIB, SSB, Fidelity Bank, Access Bank, etc. There also exist a pool of private transport owners for provision of transport services to farmers and traders as well as carpenters for manufacture and maintenance of boxes for carrying tomatoes. Financial and non-financial NGOs operating in the cluster also provide some critical services including extension and credit to actors along the tomato value chain. Many agro-inputs companies such as YARA, Wienco and Dizengoff operate in this cluster. Northern Star Tomato Company although currently not operating could play critical roles in improving the performance of the tomato industry in this cluster because of proximity. NHRC (outside the cluster), SARI, UDS, DAC and Tamale Technical University are also key players.

3.2.5 *Techiman Area Tomato Cluster*

This cluster is made up of farmers in Techiman, Nkoranzah, Wenchi, Tano and Kintampo areas. The cluster has an estimated number of over 25,000 farmers who engage in both rain-fed and irrigated vegetable production. Vegetables cultivated by most farmers include tomatoes, okra, pepper and leafy vegetables such as cabbage. Most of the farmers engage in rain-fed tomato production and cultivate less than one hectare per production cycle with total area under cultivation estimated to be about 12,000 ha. Annually, farmers in this cluster produce about 116,000 metric tons of tomatoes. There is potential for this volume of production to be increased given the necessary productivity improvements support and market linkages. The key service providers in this cluster are the Ghana Irrigation Development Authority (GIDA) for the provision of irrigation, MOFA for the provision of extension services and private individual tractor operators for land preparation services.

There is a vast array of financial services providers in this cluster including the GCB Ltd, ADB, Barclays Bank, NIB, SSB, Fidelity Bank, Access Bank, etc. There also exist a pool of private transport owners for provision of transport services to farmers and traders as well as carpenters for manufacture and maintenance of boxes for carrying tomatoes. Financial and non-financial NGOs operating in the cluster also provide some critical services including extension and credit to actors along the tomato value chain. Many agro-inputs companies such as YARA, Wienco and Dizengoff operate in use this cluster. The other agro-input dealers are Agyaaku Farms Training Ltd, Woffa Addo AgroChemicals, Frankatson Ltd, Donewell Agro-Ventures, etc., also operate in the cluster. Although currently not in operations, the Wenchi Tomato Factory (TAMACAN) and the Techiman Food Processing Complex (TEPCO) could be key players in this cluster. The University of Renewable Natural Resources (URNR), the Methodist University College Ghana (MUCG), Wenchi Campus, the Kwame Nkrumah University of Science and Technology (KNUST—outside the cluster), University of Education, Mampong Campus (UEW—outside the cluster), UDS (outside the cluster), Kwadaso Agricultural College (KAC), Sunyani Technical University, University of Energy and Natural Resources (UENR) and the Crops Research Institute (CRI), among others are key players in this cluster.

3.2.6 Akomadan Area Tomato Cluster

This cluster is made up of farmers in Offinso South Municipality and North district. The cluster has an estimated number of over 10,000 farmers who engage in both rain-fed and irrigated vegetable production. Vegetables cultivated by most farmers include tomatoes, okra, pepper and leafy vegetables such as cabbage. Most of the farmers engage in rain-fed tomato production and cultivate between half and two hectares per production cycle with total area under cultivation estimated to be about 10,000 ha. Annually, farmers in this cluster produce about 100,000 metric tons of tomatoes. There is potential for this volume of production to be increased given the necessary productivity improvements support and market linkages.

The key service providers in this cluster are the Ghana Irrigation Development Authority (GIDA) for the provision of irrigation, MOFA for the provision of extension services and private individual tractor operators for land preparation services. There is a vast array of financial services providers in this cluster including the Offinso Rural Bank Ltd, Nkwabingya Rural Bank, etc. Financial and non-financial NGOs operating in the cluster also provide some critical services including extension and credit to actors along the tomato value chain. There also exist a pool of private transport owners for provision of transport services to farmers and traders as well as carpenters for manufacture and maintenance of boxes for carrying tomatoes. The agro-key input dealers operate in nearby Techiman area cluster and these include Enepa ventures, OLAM, LDC, Sefa & J, K. Badu enterprise. Although currently not in operations, the Wenchi Tomato Factory (TAMACAN) and the Techiman Food Processing Complex (TEPCO) could be key players in this cluster given their proximity. The URNR, MUCG, KNUST, UEW, UDS (outside the cluster), KAC, Kumasi Technical University (KuTU) and CRI, among others are key players in this cluster.

3.2.7 Accra Area Tomato Cluster

This cluster is made up of farmers in Accra Metropolis, Dangme East/West, Ga East/West and Tema areas. The cluster has an estimated number of over 30,000 farmers who engage in both rain-fed and irrigated vegetable production. Vegetables cultivated by most farmers include tomatoes, onions, okra, cucumber, green pepper, chili pepper, carrot, shallot and leafy vegetables such as cabbage, among others. Most of the farmers engage in irrigated tomato production and cultivate less than one hectare per production cycle with total area under cultivation estimated to be about 15,000 ha. Annually, farmers in this cluster produce about 58,000 metric tons of tomatoes. There is potential for this volume of production to be increased if the necessary productivity improvements support and market linkages are put in place. The key service providers in this cluster are the Ghana Irrigation Development Authority (GIDA) for the provision of irrigation, MOFA for the provision of extension services and private individual tractor operators for land preparation services.

There is a vast array of financial services providers in this cluster including the GCB Ltd, ADB, Barclays Bank, NIB, SSB, GN Bank, Fidelity Bank, Access Bank, UniBank, Multicredit, Procredit, etc. Financial and non-financial NGOs operating in the cluster also provide some critical services including extension and credit to actors along the tomato value chain. There also exist a pool of private transport owners for provision of transport services to farmers and traders as well as carpenters for manufacture and maintenance of boxes for carrying tomatoes. The agro-key input dealers operate in nearby Techiman area cluster and these include Dizengoff, OLAM, Wienco, etc. The University of Ghana (UG),

Accra Technical University (ATU), Council for Scientific and Industrial Research (CSIR), GSA, FDA, EDAIF, etc., are key players here.

3.2.8 Akatsi Area Tomato Cluster

This cluster is made up of farmers in Akatsi, Keta, Ketu, Tongu, Adidome, Sogakofe and Aflao areas. The cluster has an estimated number of over 15,000 farmers who engage in both rain-fed and irrigated vegetable production. Vegetables cultivated by most farmers include tomatoes, okra, shallot, cucumber, chili pepper, green pepper and leafy vegetables. Most of the farmers engage in irrigated tomato production and cultivate between half and one hectare per production cycle with total area under cultivation estimated to be about 20,000 ha. Annually, farmers in this cluster produce about 65,000 metric tons of tomatoes. There is potential for this volume of production to be increased given the necessary productivity improvements support and market linkages.

The key service providers in this cluster are MOFA for the provision of extension services. Financial and non-financial NGOs operating in the cluster also provide some critical services including extension and credit to actors along the tomato value chain. There exist financial services providers in this cluster including the GCB, ADB, etc. There also exist a pool of private transport owners for provision of transport services to farmers and traders as well as carpenters for manufacture and maintenance of boxes for carrying tomatoes. The key agro-input dealers operating in this cluster include Dizengoff, OLAM, Wienco, etc. Tiptop Tomafresh, a small-scale tomato processing company also exists in this cluster and is a key player in the tomato value chain. The University of Health and Allied Sciences (UHAS) and Ho Technical University (HTU), among others are also key players here.

3.2.9 Ho Area Tomato Cluster

This cluster is made up of farmers in Ho Municipality, Ho West, Adaklu and Agotime-Ziope districts. The cluster has an estimated number of over 8000 farmers who engage in both rain-fed and irrigated vegetable production. Vegetables cultivated by most farmers include tomatoes, okra, pepper and leafy vegetables. Most of the farmers engage in rain-fed and irrigated tomato production and cultivate less than one hectare per production cycle with total area under cultivation estimated to be about

6000 ha. Annually, farmers in this cluster produce about 75,000 metric tons of tomatoes. There is potential for this volume of production to be increased given the necessary productivity improvements support and market linkages. The key service providers in this cluster are MOFA for the provision of extension services. Financial and non-financial NGOs operating in the cluster also provide some critical services including extension and credit to actors along the tomato value chain. There exist financial services providers in this cluster including GCB, ADB, NIB, etc. A pool of private transport owners for provision of transport services to farmers and traders as well as carpenters for manufacture and maintenance of boxes for carrying tomatoes also exists. The key agro-input dealers operating in this cluster include Dizengoff, OLAM, Wienco, etc. UAHS and HTU are also key players here.

3.2.10 Begoro Area Tomato Cluster

This cluster is made up of the East Akim Municipality, Fanteakwa, Kwahu South, Lower Manya Krobo and the Yilo Krobo Districts. The cluster has an estimated number of over 20,000 farmers who engage in vegetable production. Vegetables cultivated by most farmers include tomatoes, okra and leafy vegetables such as cabbage, among others. Most of the farmers cultivate less than one hectare per production cycle with total area under cultivation estimated to be about 10,000 ha. Annually, farmers in this cluster produce about 10,000 metric tons of tomatoes. There is potential for this volume of production to be increased if the necessary productivity improvements support and market linkages are put in place.

The key service providers in this cluster are MOFA for the provision of extension services. Financial and non-financial NGOs operating in the cluster also provide some critical services including extension and credit to actors along the tomato value chain. There exist financial services providers in this cluster including the GCB Ltd, ADB, etc. There also exist a pool of private transport owners for provision of transport services to farmers and traders as well as carpenters for manufacture and maintenance of boxes for carrying tomatoes. The key agro-input dealers operating in this cluster include Dizengoff, OLAM, Wienco, etc. The Presbyterian University College (PUC) and Koforidua Technical University (KoTU) are also key players here.

3.3 Product Identification, Domestic Capacity and Markets—Ghana Case Study

3.3.1 Product Identification

Fresh tomato for supermarkets and small-scale factories was identified as the key product to focus on during consultative meetings with farmers and other stakeholders in the tomato industry across the country. The main reason given for the choice of fresh tomatoes by stakeholders is that there is a huge domestic demand that needs to be satisfied. It came out clearly during the consultative meetings that there are competition among open-air markets, supermarkets and tomato processing factories for fresh tomatoes during some periods of the year. This competition creates a situation of uncertainty in the tomato market as open-air market 'queens' always employ manipulative strategies to outwit other buyers by offering higher prices to farmers within a very short window just to scare off the other buyers. As soon as the other buyers (i.e. factories and supermarkets) leave the scene because of the high prices, the market 'queens' quickly drop prices to their advantage. This according to stakeholders is inimical to the advancement of the tomato industry. To help improve the industry, all year-round production of tomato fresh tomatoes and this will provide the opportunity for imports substitution and wealth creation through a sustained supply of adequate raw materials to the processing companies.

3.3.2 Domestic Capacity

There is capacity for Ghana to become self-reliant in both fresh and processed tomato products. The country has a vast agricultural land suitable for tomato production and huge population of farmers willing to engage in tomato production should the right interventions be put in place. Besides, there are research institutions such as the Council for Scientific and Industrial Research (CSIR) and the many universities that have the capacity to come out with the appropriate tomato varieties and production technologies that better suit the local environment and weather conditions. For instance, the Forest and Horticultural Crops Research Centre (FOHCREC) of the University of Ghana in collaboration with Stevicksen Ventures Ltd (SVL) and Eisenberg-Agri (Beijing) Ltd have developed an Enviro Dome Greenhouse System for vegetable production (See more at: http://www.myjoyonline.com/business/

2014/April-7th/new-technology-for-vegetable-production-introduced-in-ghana.php#sthash.FfaGCSLq.dpuf). Scientists at the Crops Research Institute (CRI) and the Savannah Agricultural Research Institute (SARI) have been leading the breeding and releasing of new tomato varieties to farmers but their efforts are not adequately backed by sustained research funding. The efforts however, demonstrate the capacity of institutions to come out with the right technologies for improved productivity in the sector.

There are also tomato-processing factories of varying installed capacities dotted across the breadth and length of Ghana, which when well resourced, will be able to process tomatoes to meet domestic demand and even export. There are private individuals and companies in Ghana that import tomato pastes, package locally and distribute to various outlets across the country for onward supply to consumers. This means that the capacity to package, brand and market processed tomatoes exists locally. There are many financial institutions including universal banks, microfinance institutions, rural and community banks across the country that have the capacity to provide wide range of financial products to clients across the different productive sectors of the Ghanaian economy.

The Ministry of Food and Agriculture (MOFA) when resourced has the capacity to provide technical advisory services to tomato farmers and input dealers to deliver appropriately. The Ministry of Trade and Industry (MOTI) and the Ghana Export Promotion Authority (GEPA) through Parliament as well as other relevant bodies also have the capacity to come out with the necessary legislative and regulatory frameworks to guide business generally and the tomato industry in particular. Indeed, the current stable democratic dispensation and independent judicial system provides impetus for public–private-partnership (PPP) arrangements that will further enhance local capacity towards the goal of self-sufficiency in tomato supply in Ghana.

3.3.3 Market Analysis

The tomato market in Ghana is made up of two main segments—the fresh tomato market segment and processed tomato market segment. The fresh tomato market segment is currently dominated by products from local production augmented by imports from neighbouring countries, particularly Burkina Faso at off-peak. The mainstream channels for fresh tomatoes are open-air markets and roadsides selling with the niche markets being supermarkets and grocery stores. Appreciable proportion of fresh tomatoes sold in the niche markets is imported. The processed tomato market is largely dominated by imports from Europe and Asia. Most processed tomatoes are imported as pastes and canned or packaged in Ghana for distribution. Prices of fresh tomatoes are very volatile due to seasonality as most production systems depend on the natural climatic conditions—temperature, light and rainfall either on rain-fed or irrigated open fields. Modern production systems such as greenhouses are less developed and rarely used by tomato farmers in Ghana. Tomato prices are usually low during the rainy season, the peak production period (i.e. from July to November) and high in the dry season (i.e. December to June) each year. Most institutional consumers such as schools and hospitals prefer processed tomatoes because of convenience and stability in supply all year-round which is important for budgeting and planning. Most households however, prefer fresh to processed tomatoes because of the belief that the former is healthier than the later. Currently, the processed tomato market is competitive as there is free entry and exit by actors. The fresh tomato market is however, governed by 'market queens' who somehow control entry. Key product requirements by customers include low agrochemical use and long shelf life.

3.3.4 Profitability Analyses of Irrigated, Rain-Fed and Greenhouse Tomato Production

The analyses in Tables 3.1 and 3.2 indicate that investment in tomato production is financially profitable. There is a quick turn over and this should encourage entrepreneurs to invest in this sector. In addition to labour expenditures as a share of total production cost, which is about 45% for irrigated and about 47% for rain-fed, a significant share of production cost in Ghana is spent on fertilizer and manure (32% for irrigated and 33% for rain-fed). Minimizing overall production costs by increasing the labour force productivity and increasing yields (by using the biological potential of the hybrids with proper fertilization, introducing soil testing, etc.) would increase productivity of the value chain. The huge initial capital investment required for greenhouse production presents a challenge but the returns are equally high that presents a lucrative business opportunity to be had. For detailed analyses of the potential of greenhouse tomato production in Ghana and its profitability, refer to the

Table 3.1 Estimated cost of tomato production per hectare

Estimates for 2016 cropping season			
Activity	**Cost (GH¢) Irrigable**	**Cost (GH¢) Rain-Fed**	**Cost (GH¢) Greenhouse**
Land rent	400	400	400
Land preparation	750	750	3000
Seeds (100g x 4)	360	360	720
Fertilizer & manure	4650	4650	18,600
Agro-chemicals	300	300	1200
Watering	500	0	2000
Staking	400	400	1600
Mulching	600	600	2400
Labour	6500	6500	26,000
Cost of Greenhouses	-	-	2,300,000
Estimated loan amount	14,230	13,960	2,355,920
Loan duration (Years)	1	1	2
Interest at 30% p.a	4269	4188	1,406,752
Total estimated cost	18,499	18,148	3,762,672
Interest for year 1	4269	4188	703,376
Payment due in year 1	18,499	18,148	1,881,336
Estimates for 2017 cropping season			
Activity	**Cost (GH¢) Irrigable**	**Cost (GH¢) Rain-Fed**	**Cost (GH¢) Greenhouse**
Land rent	450	450	450
Land preparation	800	800	3200
Seeds (100 g × 4)	400	400	800
Fertilizer and manure	5000	5000	20,000
Agrochemicals	350	350	1400
Watering	550	0	2400
Staking	450	450	1800
Mulching	600	600	2400
Labour	7500	7500	30,000
Greenhouses maintenance	–	–	50,000
Estimated loan amount	16,100	15,550	112,450
Loan duration (Years)	1	1	1
Interest at 32% p.a	5152	4976	35,984
Total estimated cost	21,252	20,526	148,434
Outstanding loan (Yr 2)	21,252	20,526	148,434
Payment due in year 2	21,252	20,526	2,029,770

Table 3.2 Estimated revenue from a hectare of tomato production

Estimates for 2016 cropping season			
Item	**Irrigable**	**Rain-Fed**	**Greenhouse**
Yield (Kg)	35,000	30,000	4,000,000
Percentage loss	10	10	10
Available yield (Kg)	31,500	27,000	3,600,000
Packaging in boxes (52 kg)	605.8	519.2	69,230.8
Farm gate price (GHS/52 kg)	50	50	50
Gross income (GH¢)	30,290	25,960	3,461,540
Net income (Revenue—cost)	11,791	7812	1,580,204
Outstanding loan in year 2	0	0	1,881,336
Estimates for 2017 cropping season			
Yield (Kg)	35,000	30,000	4,000,000
Percentage loss	10	10	10
Available yield (Kg)	31,500	27,000	3,600,000
Packaging in boxes (52 kg)	605.8	519.2	69,230.8
Farm gate price (GHS/52 kg)	55	55	55
Gross income (GH¢)	33,319	28,556	3,807,694
Net income (Revenue—cost)	12,067	8030	1,777,924
Outstanding loan in year 2	0	0	0
Debt Service Ratio (DSR)	64%	72%	53%

comprehensive study on strategies to support the greenhouse horticulture sector in Ghana by Elings, Saavedra and Nkansah (2014).

3.4 The Action Arenas—The Case of Tomato Production in Ghana

3.4.1 The Inputs Arena

This arena is made up of labour, seeds and agrochemicals suppliers. Family labour is mostly employed in the production of tomatoes. Bulk of the labour is used during the most labour intensive activities like planting, weeding, hoeing and transporting. Farmers use seed from their own produced tomatoes (i.e. washed seed) or buy from input dealers (i.e. improved seeds). The commonly sold seed varieties are Petomech, tropimech, UC, No-name, Power Rhinno and Rasta. Most of the input dealers only sell agrochemicals with a few of them selling tomato seeds. Generally, the agrochemicals used in tomato production are fertilizers, pesticides/insecticides and weedicides. The main actors in this arena are

Dizengoff Ghana, Wienco, Agrimat and Yara with a lot of distributors, wholesalers and retailers across the country. On the average, wholesalers add a margin of between GH¢10 and GH¢20 per unit of the factory price and retailers in turn add between GH¢2 and GH¢10 per unit of the wholesale price depending on how far the town or community is from the wholesaler. Thus prices of inputs remain largely uniform across dealers and the only thing that makes the difference is social relationships with farmers.

Although all agro-input dealers are supposed to register with the umbrella body, Ghana Agricultural Input Dealers Association (GAIDA) there are so many unregistered dealers, especially at the retail end of the arena. Membership to the group is open to all who are known agro-input dealers and willing to pay the registration fees. In clusters such as the Navrongo area cluster, the registration fee is of about GH¢100 and monthly dues of GH¢4 thereafter. These payments are used to finance the activities of the group. The main benefits of belonging to the group are increased bargaining power with suppliers and customers and ability to get loans from financial institutions with favourable conditions. Group membership also brings about free flow of information, especially on prices among dealers. There is also an additional benefit of working together as a group to help prevent the circulation of fake chemicals. Group members also get their capacities built through training on proper chemical handling and usage so as to be able to provide advisory services to their customers, mainly farmers. Leaders of GAIDA across the different clusters do organize talks for farmers on the use of agrochemicals via radio from time to time, especially when new inputs are being introduced. One of the key challenges of GAIDA in this cluster is the refusal or untimely payment of dues by members, which negatively affects the capacity building activities of the group. Also, most input dealers at the retail end do not have adequate training on input handling and use although farmers rely on their advice and judgement for application of inputs. Another key challenge in this arena is the presence of counterfeit agro-inputs.

Across all clusters in the country, inputs are available all year-round as some farmers engage in dry season or irrigated farming. Thus input dealers make sure that they are supplied with enough inputs for all types of crops to prevent shortages and to also be able to stay in business. They are able to access inputs either by going to the distributing towns/cities or by the services of input suppliers distributing to dealers in their various towns when they request through phone calls and sending money through banks, mobile money or other means.

Although input dealers complained that the prices of farming inputs are not steady, they are able to maneuver funds to ensure they do not run out of stock or business, as it is their source of livelihood. The main means of funding for input dealers are savings and bank loans. Sometimes the suppliers of the inputs supply on credit. There are high fluctuations in prices due to the instability of the national currency (Table 3.3).

Table 3.3 Strengths, weaknesses, opportunities and threats (SWOT) of the Arena in 3.4.1

Internal Factors	
Strengths	**Weaknessse**
1. Readiness of entrepreneurs to invest in agro-inputs enterprises	1. Low financial capacity and other operating resources
2. Willingness of GAIDA and its members to work for development of the inputs arena	2. Low skills levels among some input dealers, especially at the retail end
3. GAIDA as a business entity is independent of the bureaucracies of the mainstream institutional system	
External Factors	
Opportunities	**Threats**
1. There exist a huge customer base across the country for agro-inputs	1. Low level of appreciation and orientation of agribusiness as central to rural development
2. There exists a critical mass of institutions with interest in agribusiness for collaboration	2. Presence of unregistered dealers most of who engage in sales of fake agro-inputs to farmers
3. A critical mass of people, especially women and youth, needing training to become job creators in the agribusiness sector exists	3. Instability of the national currency causes fluctuations in product prices
4. Stable political and economic environment for businesses in Ghana.	4. Low level of relevant agricultural technological innovativeness
5. Agriculture is visibly positioned on the political agenda	5. High cost of bank transactions and general reluctance of financial institutions to fund agribusinesses
6. There are several organizations and foundations worldwide that are willing to support agribusiness training and agribusinesses	6. Poor rainfall patterns that affect demand for agro-inputs

3.4.2 The Services Arena

This arena is made up of different actors who provide critical services for the production of tomatoes and these include agricultural extension services, mechanization services, financial services, irrigation services as well as transport and logistics services. MOFA through its extension agents provide extension services although these services are very general and not tailored to tomato production. GIDA also provides some water services to farmers. Market queens and traders provide marketing services. Private individuals provide mechanization services using tractors and animal-drawn implements. The use of tractors for land preparation is important especially in the open-field tomato production, as farmlands need to be ploughed first before sowing or planting can be done. Tractors are not readily available in most clusters, especially at peak times and this affects the production process. The few tractor operators always make themselves available to farmers by working longer hours, typically from 4am to 10 pm to the extent that they hardly rest or spend time with their families. Tractor operators charge between GH¢50 and GH¢100 per acre depending on the nature of the land (i.e. hard or soft, free of trees or not, clayey or sandy very wet or not). Actors providing mechanization services are not in any organized group.

There are a number of financial service providers across the different tomato clusters providing financial services for actors along the value chain. These include bank and non-bank financial institutions. The bank financial institutions with presence across the country include the GCB Bank Ltd, Agricultural Development Bank Ltd, Rural and Community Banks, and the non-bank financial institutions are largely micro-finance institutions (MFIs), credit unions and financial NGOs. There also exist the informal financial actors such as moneylenders and Rotation Savings and Credit Associations (ROSCAs) that provide financial services to actors along the value chain. These financial institutions provide various forms of financial services to the different actors within the tomato value chain. They provide credit facilities, especially to the input dealers and traders. Unfortunately, most of these financial service providers are unwilling to provide credit to tomato producers as a result of high risks associated with that segment of the value chain, exacerbated by the lack of comprehensive insurance policy for crop farmers.

Transporters render critical services to the producers, input dealers and marketers. Members of the Ghana Private Road Transport Union

(GPRTU) of the Trade Union Congress (TUC) and other transport unions and individuals as well as logistics companies provide essential services to the tomato value chain. Transporters are the actors that move inputs from the manufacturing points to the farm and in turn carry the produce from the farm gate to the end markets and consumers. Other service providers (Table 3.4).

Table 3.4 The services Arena SWOT table in 3.4.2

Internal Factors	
Strengths	**Weaknessse**
1. Presence of service providers (e.g. banks, MOFA, transporters) across the country 2. High demand for services 3. Presence of strong local transporters union (GPRTU) 4. Willingness of entrepreneurs to invest in the transport industry 5. Availability of spare parts and local human resource for repairs and management of vehicles	1. Unwillingness of some farmers to pay for services (e.g. extension) 2. Low financial capacity and other operating resources 3. Low level of tomato production skills among farmers, especially women and youth
External Factors	
Opportunities	**Threats**
1. Increasing demand for transport services due to urbanization and economic development 2. There exists a critical mass of entrepreneurs willing to invest in this arena 3. A critical mass of people, especially women and youth, needing training to become job creators in the agribusiness sector exists 4. Stable political and economic environment for businesses in Ghana 5. Agriculture is visibly positioned on the political agenda	1. Poor road networks in production areas 2. High prices of spare parts 3. Frequent road accidents 4. Refusal of some transporters to join organized associations 5. Instability in fuel prices occasioned by volatility in exchange rates

3.4.3 The Production Arena

This arena is made up mainly producers of different scales of operations. Farmers in this cluster engage in both rain-fed and irrigated production systems. Those into irrigated production system rely on irrigation dams and natural water bodies such as rivers. Smallholder farmers dominate this arena with farm holdings less than 2 hectares per farmer. Yields are also low due to a gamut of factors including high incidences of pests and diseases, poor rainfall patterns, high night temperatures, use of low-quality seeds, poor soil fertility, etc.

The common types of tomatoes produced by farmers include Petomech, tropimech, UC, No-name, Power Rhinno, Rasta and a local variety, which is a mixture of various varieties. Some farmers have started using washed seeds from the Burkina tomato and complained that the variety keeps changing after every harvest.

Farmers lack collective bargaining and marketing power because they do not belong to a well functioning group and the free market system, which allows tomato imports from other countries such as Burkina Faso. Although the farmers were part of a group, they had never come together for the marketing of their tomatoes. Farmers sell their produce individually and immediately after harvest since tomatoes are highly perishable. Farmers usually transport their produce to the market themselves using different means of transport such as taxis, 'trotro', trucks and motorcycles/'motor kings'. They readily sell to the general public, which includes individuals, wholesalers and retailers.

The agricultural inputs used by farmers for tomato production are fertilizers (Urea, NPK and Sulphate of Ammonia) and chemicals such as top-up (also serves a folia fertilizer being absorbed through the leaves), furadan, which protects against nematodes, lambda for spraying when there are pests and miktine for unfavorable weather conditions. They also use chemicals like, harvest more, grow force, super force (used during the flowering stage to kill the larvae of insects) and cold size. From our interactions, it shows that the farmers obtain most of their inputs from the shops in the clusters within which they operate. Some of these shops sell multiple inputs (like seed, fertilizers and agrochemicals) while others sell only one input (like fertilizer or chemicals). Farmers are located very close to the input markets and thus the availability of seeds and other inputs at crucial planting time is not an issue. However, there was the issue of increases in prices on one hand and the flow of inferior quality inputs on the other. Farmers say that due to inadequate funds they buy agrochemicals based on how much they can afford and not the quantity needed for their farm size.

Farmers are afraid of taking loans or credit to invest in their tomato production since tomatoes farming is risky and one cannot determine when there will be ready market for their produce or when they will poor harvest due to bad weather. Some farmers however, turn to selling of livestock, crops and other properties to be able to buy farm inputs for production.

Farmers need training on correct package of practices and application of inputs. Some tomato farmers do not stake but only do mulching, which they use crop residue particularly rice straw, which is not bought but conveyed by them from rice fields. It can take 1 day to apply mulch on an acre of land. The farmers themselves do the marketing of their produce. Expenses for buying seeds, fertilizers, chemicals, fuel for irrigating the land, labour, etc., all come from the farmer's pocket. Some interesting facts emerged when farmers started talking about different expenses in growing tomatoes. For instance, the farmers did not consider their own work on the farm as an expense. They only consider labour as an expense when they have to employ/hire people or somebody, popularly known as farm boys and pay them. There is lack of trust between input suppliers and farmers. Farmers believe that input suppliers do not provide good quality inputs while the inputs suppliers say that farmers do not have adequate skills to use the inputs correctly. The input suppliers also say that they are only 'suppliers' and not manufacturers of inputs. In other words they sell whatever they receive from their suppliers (Table 3.5).

3.4.4 Output Marketing Arena

This arena is made up of different actors and channels and these are:

- Brokers/Aggregators
- Wholesalers
- Open-air/roadside retailers
- Supermarket channels
- Small grocery stores
- Processing companies
- Institutional consumers

Table 3.5 Strengths, weaknesses, opportunities and threats (SWOT) of the Arena in 3.4.3

Internal Factors	
Strengths 1. Energetic men, women and youth that can apply knowledge and technology for tomato production 2. Willingness of farmers to commit themselves to work for development of the tomato value chain	**Weaknessse** 1. Low financial capacity and other operating resources 2. Low level of tomato production skills among farmers, especially women and youth
External Factors	
Opportunities 1. Increasing rate of urbanization and changes in consumer tastes raises domestic demand for tomato and tomato products 2. There exists a critical mass of institutions with interest in agribusiness for collaboration. 3. A critical mass of people, especially women and youth, needing the training to become job creators in the agribusiness sector exists 4. Stable political and economic environment for businesses in Ghana 5. Agriculture is visibly positioned on the political agenda	**Threats** 1. Low image of farming and agriculture-related businesses among the youth and educated segment of the Ghanaian population 2. Low level of appreciation and orientation of agribusiness as central to rural development by many people including policymakers, political leaders and even some 'development experts' 3. Low standards of product development by actors in the agricultural value chain 4. High dumping of cheap foreign tomato products in the Ghanaian market 5. Low level of relevant agricultural technological innovativeness 6. High cost of bank transactions and general reluctance of financial institutions to fund farmers

a. **Brokers/Aggregators**

The role of brokers is central to the bulk of vegetables sales cross Ghana as a lot of tomatoes pass through them. They act as a gateway of information between the wholesales, small to medium traders, supermarkets and the producers. They physically do not add any value to the product; they basically mediate the transactions between the producers and the buyers. This channel is particularly popular with supermarkets. The brokers/aggregators have well-established network of buyers and sellers. The process usually is initiated by the big buyers in markets in the cities such as Kumasi, Accra, Tamale, etc. The buyers raise their demands with the brokers specifying the quantity and quality requirements. Using the information on quality and quantity requirements, the brokers go to look for the availability of the required products from farmers. At times they collect samples from the farmers to show it to the buyers before negotiating the price with sellers. Thus the brokers mediate the transactions between the producers and buyers. For instance if the total price to be paid by the buyers is GH¢1000 then an advance of say GH¢400 is paid, usually a day before the delivery, to the farmer by the buyer and the balance is paid at the time of delivery. During this time the buyer is not usually present and all the transactions happen through the broker. In other words the broker takes the responsibility of ensuring that the money reaches the farmers and the buyers get the product. The brokers in turn supply the material to the wholesalers in the big markets.

b. **Wholesalers**

The wholesalers buy the bulk of the produce from farmers and in some cases brokers/aggregators and supply the product to retailers and traders. Wholesalers always prefer to buy the product from farmers because of the belief that brokers offer tomatoes at high prices, making good margins and that affect the profit margins of wholesalers. Thus wholesalers sometimes go into verbal contracts with farmers. Unfortunately, some wholesalers have had some bad experiences where the farmers did not adhere to the verbal agreements on quantity, quality and timeliness. With slight positive change in market prices of tomatoes, farmers most likely shift from one buyer to the other going back on their agreements.

c. **Open-air/roadside retailers**

This market channel currently dominates the retail channel for fresh tomatoes and even processed tomato products in Ghana in terms of sales volume. It is estimated that about 80% of fresh tomato produce in Ghana passes through this market channel. Open-air markets handle large volumes of produce, hence competition is high and sale prices are low compared to other channels. Quality requirements for tomatoes in this channel are not sophisticated and competition is mainly based on price. Consumers in this segment choose the product that they feel offer the best value proposition (price in relation to appearance). Transactions in this channel among producers, traders and consumers are mostly informal, with no invoices or receipts used. Significant issue also is the lack of enforcement of sanitary and phytosanitary standards. Consumers prefer to purchase tomatoes in this channel due to an overall perception that the produce here is better quality and fresher than in supermarkets where produce stays on the shelf for a longer time. At the height of the production season, temporary wholesale open-air markets are established in production centres where local producers sell their produce to traders. At the local open-air markets, growers sell their produce to aggregators/wholesalers who in turn transport it to the major market centres (e.g. Techiman, Kumasi, Accra, Tamale). The key actors in this channel are the farmers, market queens/traders, transporters, lead/loading boys and consumers.

d. **Supermarkets**

The major supermarket chains in Ghana are situated in the cities, particularly Accra and include Shoprite, Game, Koala, Marina, Palace and others. Although fruit and vegetables occupy small vending spaces within typical Ghanaian supermarkets, the channel has been growing in recent times as food sales outlet and is projected to continue to grow. Supermarkets sell both local and imported fresh tomatoes and tomato-based processed products. It must be noted that this channel is the most regulated one in terms of quality, sanitary standards and supplier requirements. So the capacity of local producers must be built to be able to meet the volume requirements, delivery schedules and product standards established by the supermarkets. Thus farmers need to improve their product quality and consistency of production volumes and establish relationships with brokers and traders that supply supermarkets. This is critical because supermarkets are looking for consistent supply from a limited number of trusted suppliers.

e. **Small grocery stores**

This channel includes small local grocery/convenience stores and kiosks that trade smaller volumes and offer less variety at prices marginally higher than the open-air retail markets. This channel serves consumers who are looking to purchase a mix of different food products and beverages in one nearby location, in addition to fresh produce. Many of these stores do not offer a full range of fruits and vegetables, which occupy small proportions of their total shelf space. The key actors in this channel will be identified and their functions quantified and valued.

f. **Processing Companies**

Processing companies such as Northern Star and Wenchi tomato factories among others are key actors in this channel. Processors are expected to be collecting fresh tomatoes from farmers and aggregators for processing. Unfortunately, none of the large processing companies is currently operating because of inadequate availability of raw materials all year-round. Tomato production in Ghana must therefore be structured in such a way that processing companies will not experience difficulty getting sufficient supply from growers on a consistent basis. Indeed, processors might need to sign forward contracts with farmers and farmer-based organizations (FBOs). To make such contracts work, efforts to ensure that farmers stop side selling must be implemented. This requires that a basic amount be agreed upon before harvest with farmers and buyers agreeing to share the difference in price equally during harvest. Thus, if prices at harvest are higher than agreed, farmers should get up to at least 50% of the difference in addition to the base price and the vice versa. For instance, if the agreed price is GH¢10 and the price at time of harvest is GH¢12, then, the farmer and the buyer should the difference of GH¢2. Thus the farmer gets GH¢11. On the other hand, if at the time of harvest the price is GH¢8, then, the farmer gets GH¢9.

g. **Institutional consumers**

There are so many institutional consumers of tomatoes in Ghana including schools, hospitals, restaurants and hotels, among others. Consumers in this market segment demand high standards and consistent supplies. Thus the capacity of tomato producers should be built so that they could produce to meet the sanitary and phytosanitary standards as a primary requirement in this market segment. Besides, farmers need to establish relationships with buyers in this market for long-term business deals.

3.5 Value Addition, Vertical and Horizontal Linkages—The Tomato Case in Ghana

3.5.1 Analysis of Value Added

To understand the power dynamics within the value chain it is important to analyse the value distribution. Due to seasonality of tomato production the price of fresh tomatoes varies throughout the year. In the dry season retail price for fresh tomatoes can be as high as GH¢500/52 kg and in August/September around GH¢50/52 kg. Margins across the different segments differ significantly. On the average price margins vary between GH¢10 and GH¢50/52 kg at each level and this usually is sufficient for each of the actors to cover their costs and generate profit. A retailer would usually receive a margin of GH¢5–10/52 kg minimum in high season. At the end of the season, starting from beginning of October retailer margins may go up to GH¢5–20/52 kg. The analysis per 52 kg crate shows that the different actors along the chain add different levels of value. Producers add a total value of GH¢1230, wholesalers add about GH¢230 and retailers add about GH¢150. Overall, from producers to retailers a total value of GH¢1610 is added. The value added is distributed as follows: 76.4% by farmers, 14.3% by wholesalers and 9.3% by retailers. Thus producers make a net profit of GH¢61.75, wholesalers make a net profit of GH¢110 and retailers make a net profit of GH¢10. The wholesalers make the highest margin (61%), followed by producers (34%) and retailers make the least margin (5%). This means that the distribution of value generated in the chain is not equitable.

3.5.2 Vertical Linkages

Farmers either sell their produce at the farm gate to traders, brokers/aggregators or choose to take their produce to either the wholesale or retail market and trade directly. Long-term working relationships between farmers and intermediaries are common, and transactions are market and relationship based with competition based on price. Producer bargaining power vis-a-vis traders is limited by the number of traders in the market. This makes competition in the market quite volatile and unpredictable. Often growers representing themselves in an open-air market, having to farm still would dump the prices at the day's end in order just to leave since other business is waiting for them in their farms.

The next day it is quite difficult for them to raise back up the prices because the customers have already perceived a lower price and would expect it to be low when they will buy the next time.

Transactions between growers and traders are sporadic with a low cost of switching from producer to producer. Producers get different prices for their tomatoes depending on which retail outlet the produce is ultimately sold. Due to high risks, producers frequently sell their produce below the market price and this leads to reduction in the market price for other farmers. With limited financial analysis skills, farmers frequently do not understand the impact of the loss on their overall profitability and do not see the value in marketing in groups or through functional farmer-based organizations (FBOs).

Transfer of market information and requirements to growers is also limited by their lack of direct contact with final consumers at the high-value end market. Farmers who are able to trade directly at retail markets do so in the open-air markets where quality requirements are low and transactions are informal, so there is no incentive for them to perform well on quality. For the purposes of selling to supermarkets farmers must try to increase their quality and consistency in supply. At the same time, lack of capacity to supply consistent volumes of produce, poor quality and lack of certifications limit producer access to brokers/aggregators and traders who supply to higher value retail channels such as supermarkets, and are interested in developing longer-term relationships with growers to ensure consistency of supply. Some traders interviewed for this study indicated they might consider establishing contract relationships with growers to ensure consistent supply. This is not the main option for sourcing tomatoes because of poor production practices and seasonality, which greatly affects farmers' ability to supply the necessary volumes of tomatoes. In addition, due to lack of long-term strategic planning, farmers frequently side-sell and do not fulfil contracts if they see an immediate cash opportunity, often at a long-term loss for themselves. As a result, traders are not so willing to provide the required amount of inputs (seeds and fertilizer) to producers to ensure the needed volume of supply. Some traders had reservations about poor production practices (such as lack of irrigation, appropriate fertilization, pest management, etc.) that mostly make farmers unable to get the yields expected by the traders and processors for that matter.

To increase the flow of learning and benefits to growers, uncertainty and risks for both growers and traders need to be reduced by making the vertical relationships within the value chain more balanced and interdependent. For example, permanent-working relationships between growers and intermediaries would result in lower overall price fluctuation for the growers and reduced risks. Opportunities exist to increase supply to supermarkets, as a channel with a long-term growth trajectory, through establishing seasonal contracts with brokers/aggregators who supply supermarkets. Such arrangements would reduce the frequency of negotiation for growers, limiting market uncertainty and would ensure a price and a market outlet for a share or all of the harvest. Farmers would limit their exposure to price fluctuations of the open market, reduce costs of transporting produce to market and increase their bargaining power overall. In addition, in the long-term this will increase the flow of market information to farmers and will create incentives for improved quality.

In addition to building relationships with traders or brokers/aggregators supplying supermarkets, the production season also needs to be extended to ensure the needed volumes of supply for a longer period. According to supermarkets interviewed for this study if local producers and processors could ensure volumes and quality of supply they will not rely on imports. Currently, farmers expect payment on the spot and do not understand the profitability of such business models working with supermarkets (including invoices, paper trails and sometimes post payment), especially as growers do not typically have developed cash flows or bank accounts. Farmers prefer cash believing that they do not have to pay taxes on their earnings to the government, which is making their life more difficult than it already is.

3.5.3 Horizontal Linkages

Horizontal farmer collaborations are currently limited by lack of a commercial mentality of farmers, lack of robust farmer cooperation and investments. Joint marketing and other horizontal collaboration among tomato farmers are weak or even non-existing due to lack of overall awareness about the benefits and market opportunities such cooperation could bring. Generally, farmers do not trust any form of formal association or cooperative for marketing their products, and as a result the interests of vegetable producers at large are also not represented. A short-term mentality results in low incentives to change and accept the requirements of new business models, such as post-payment terms, which contradict the current practices of getting cash payment on the spot. Some positive examples of collaboration exist with the input dealers who belong to GAIDA which offers some services to its members such as training seminars, joint procurement of inputs (e.g. fertilizers, pesticides, seeds, etc.) and joint product marketing.

3.6 Proposed Business Models—The Case of Tomato Production in Ghana

3.6.1 Market Led Business Model

This business model (Fig. 3.1) is very much like the one currently in operation across the country and is not necessarily led by any actor. The only difference here is the attempt to forge formal relationships among actors across the different segments of the value chain. The segments in this business model are the input supplies, farmers, traders, processors and the end market segments all of which are served by the relevant service providers to be able to execute their mandated responsibilities. The model is built on formalization of win–win relationships among input suppliers–producers–traders/off-takers. In this business model, producers are strongly encouraged to sell the produce to fresh tomato traders, wholesalers/aggregators with well-equipped storage facilities and/or processors through farmer-based organizations (FBOs). Selling through FBOs is strongly recommended because it will help improve the bargaining power of producers with buyers. The FBOs should therefore act as liaisons between producers and buyers. Wholesalers could also sell to on-farm processors, industrial processors and/or fresh tomato traders. The fresh tomato traders, industrial and on-farm processors are able to supply the right quality and quantity of tomatoes and tomato products to consumers in the end markets (i.e. open-air markets, supermarkets and institutional consumers—schools, hospitals, hotels, etc.). All the actors as indicated in Fig. 3.1 are to be supported by service providers such as technical experts from GIDA, GSA, FDA, extension agents from MoFA, agricultural trade specialists from MOTI, tractor operators, financial institutions, spraying gangs, transporters/logistics companies, carpenters for the construction of boxes for tomato haulage, lead/loading boys and so on.

To ensure that the model operates smoothly, the establishment of Cluster Working Group (CWG) comprising representatives of actors

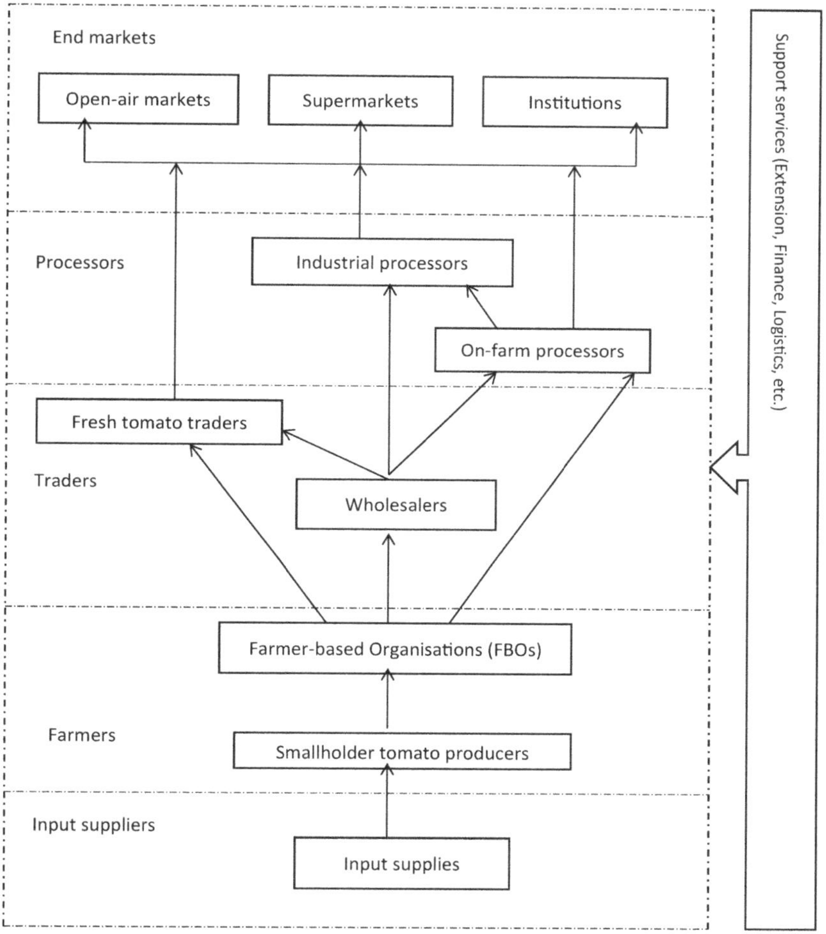

Fig. 3.1 Market led business model developed in collaboration with stakeholders

across the chain is recommended. This group has the responsibility of identifying challenges along with the different segments of the chain and proffering solutions for the collective benefit of actors. As indicated, membership of the group should come from representatives of the key segments of the value chain—inputs suppliers, farmers, traders, processors and key end market players (i.e. market queens, supermarkets and institutional buyers) and service providers (i.e. extension, finance, and logistics). All members of the working group must be operating within the cluster. In exceptional situations however, key actors located outside a cluster who are playing roles critical for the performance of the tomato industry in that cluster could be members of the working group.

3.6.2 Input Dealer Led Business Model for Fresh and Industrial Tomatoes

In this business model, an input dealer must undertake to supply all the necessary inputs for tomato production to farmers. Farmers must engage in block farming and work as producer collectives through properly instituted FBOs for easy monitoring. The key players in this business model include FBOs, recognized input dealers (e.g. Dizengoff Ghana, Wienco, YARA, etc.), MOFA, Ministry of Trade and Industry (MOTI), Export Trade, Agricultural and Industrial Development Fund (EDAIF) and its designated financial institutions (DFIs), Ghana Standards Authority (GSA) and off-takers (e.g. Shoprite, Kuala, School Feeding Programme, Senior High Schools, Colleges of Education, Northern Star, etc.). These key players must be committed to the process for it to work successfully. The input dealers must be willing to supply inputs to farmers at wholesale prices; MOFA willing to dedicate some technical staff specialized in vegetable production to provide technical services to farmers; MOTI willing to dedicate some staff specialized in agricultural commodity marketing to provide technical services to off-takers and FBOs; the EDAIF through its DFIs willing to provide funding guarantee for the input dealers to supply the inputs to farmers; GSA and GFA willing to expedite standardization and certification of tomatoes for high-value end markets; farmers willing to work in groups and to sign contracts with input suppliers, EDAIF DFIs and off-takers to abide by the terms and conditions of the contracts; and the processor and the supermarkets willing to off-take at pre-agreed conditions and to pay farmers through the EDAIF DFIs for them to redeem the funding guarantee to the input suppliers.

In summary, EDAIF provides a funding to an input dealer who upon the advice of R&D institutions on inputs requirements supplies inputs to FBOs that works with the support of the R&D institutions as well as the relevant state institutions (i.e. MOFA, MOTI, GSA &FDA) to produce industrial and fresh tomatoes for processors and supermarkets, respectively. The off-takers (i.e. processor and supermarkets) must pay for the produce through EDAIF DFIs, which will in turn deduct the costs of inputs and the rest paid to the FBOs.

To guarantee prices, contracts have to be in such a way that difference in prices at the time of harvest is shared between FBOs and off-takers using a win–win formula. Thus in the event that prices are lower, the difference between the prevailing and agreed prices should be shared in the proportions of 50% to the off-takers and 50% to the FBOs. On the other hand, if prices are higher, the difference between the prevailing and agreed prices should be shared in the proportions of 25% to the off-takers and 75% to the FBOs so as to prevent side selling by producers. For instance, if the agreed price per kilogram of tomatoes is GH¢4 and at the time of harvest the price drops to GH¢3/kg, the FBOs should be paid GH¢3.5/kg. Also, if the price at the time of harvest rises to GH¢5/kg, then, the FBOs should be paid GH¢4.75/kg. It is recommended that for the production of industrial tomatoes FBOs should use the open-field production system and irrigation facilities should be provided to ensure all year production and supply of tomatoes. For the fresh tomatoes for supermarkets, the use of greenhouse technology is recommended (Fig. 3.2).

3.6.3 Nucleus Farmer/Processor Led Business Model

In this business model, a processor must undertake to work with an input dealer that can supply all the necessary inputs for tomato production to farmers who are willing to engage in block farming through farmer-based organizations (FBOs) for easy monitoring (Fig. 3.3).

This business model is very similar to the input dealer led business model (see Sect. 3.4.2 for descriptions) with the only difference being that FBOs are only producing industrial tomatoes for the processor/nucleus farmer who is leading the process. For sustainability, the processor must also be a nucleus farmer and should be supported by EDAIF to produce up to 60% of tomatoes required for processing. To ensure that farmers produce at the least cost, the open-field tomato production system must be used in this model.

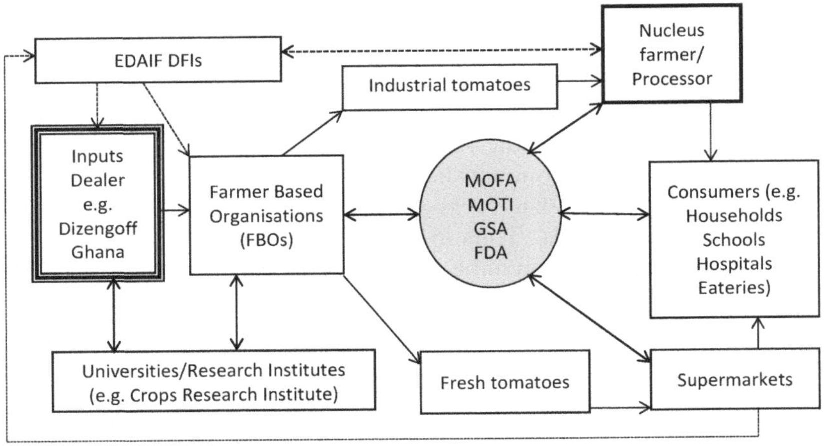

Fig. 3.2 Dealer led business model developed in collaboration with stakeholders

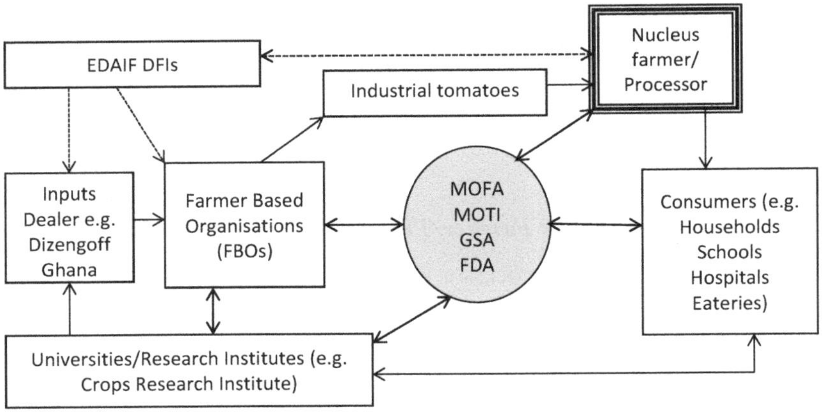

Fig. 3.3 Processor led business model developed in collaboration with stakeholders

3.6.4 *Supermarket Led Business Model*

In this business model, a supermarket must undertake to work with an input dealer and EDAIF DFIs to supply all the necessary inputs for tomato production to farmers willing to engage in block farming through farmer-based organizations (FBOs) for easy monitoring. This business model is very similar to the input dealer led business model (see Sect. 3.4.2 for descriptions) with the only difference being that FBOs are only producing fresh tomatoes for the supermarkets that is leading the process. To ensure consistent supply of tomatoes to supermarkets all year-round, greenhouses production system should be used for this model (Fig. 3.4).

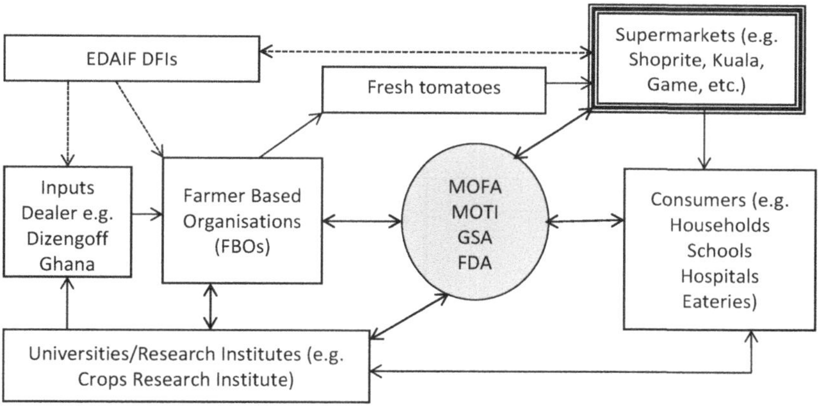

Fig. 3.4 Supermarkets led BM developed in collaboration with stakeholders

CHAPTER 4

Priority Products, Supporting Services and Institutions

Abstract Building robust commodity value chains requires prioritization and provision of the relevant support services by private and public institutions. Thus, it is important for value chain actors and stakeholders to prioritize their actions towards developing specific products or commodities that they have competitive advantage in. This is important because it makes business sense to prioritize the production of products that meet the demands, needs and preferences of the target consumers. In this chapter, we present how commodity products could be prioritized and the supporting services that are required for success. We also discuss in this chapter the regulatory institutions that must play critical roles in providing the requisite support services for the development of chosen priority products.

Keywords Commodity chains · Priority products · Regulatory institutions · Stakeholders · Support services

4.1 Priority Commodity Products

Stakeholders and value chain actors must be committed to ensuring that the performance of commodity supply and value chains they engage in is transformed by investing in the development of priority products. The

A.-R. Alhassan and M. Abunga Akudugu,
Supply Chain Management in African Agriculture,
https://doi.org/10.1007/978-3-030-54209-2_4

level of commitment of the value chain actors and stakeholders measured by their willingness to actively participate in consultative processes for performance improvement must very high for change to happen. The actors and stakeholders must clearly understand that their commodity supply and value chains are faced with numerous challenges ranging from pests and diseases to technology all underpinned by actions of political leaders. They must see the need for a coordinated approach to addressing the challenges facing specific commodity supply and value chains. Producers, traders, brokers and processors must all be willing to invest time and resources towards a self-sufficient commodity supply and value chains. They must understand the time and resource investments required at each segment of the chain, the risks therein and the type of leadership required for effective functioning of the chain for the benefit of all actors. The level of motivation of the value chain actors and stakeholders influenced by the high prospects of better returns on investments in terms of profits requires a holistic strategy through the value chain approach to deal with the challenges of the industry.

The main priority products that require intervention must be selected based on evidence and not emotions. These products should be selected for intervention based on the fact that there is high domestic demand for them and high potential for exports. For example, products that countries spend huge sums of monies each year importing them from other countries due to the fact that domestic production is unable to meet demand should qualify for priority intervention. Thus by designing and implementing the appropriate interventions to promote domestic production of these products will lift the import burden on the country. The interventions will ultimately lead to import substitution strategy resulting in long-term self-sufficiency in production in the country concern. There must also be identification of segments of the commodity supply and value chains that require priority interventions. And this should be based on the fact that the effectiveness of these segments of the chain is key in helping the country become self-reliant in the production of the particular commodity. Although the entire value chain must be considered, those identified segments must deserve special attention because of the highest risks factors, low profitability and investments levels in these segments. Infact, stakeholders identified those segments as the weakest links in the supply and value chains.

4.2 Supporting Services

4.2.1 Seed Suppliers

The supply of quality seeds is a prerequisite to successful commodity supply and value chains. However, the seed sector in Africa is generally underdeveloped with informal seed producers dominating the seed arena. Most efforts to ensure the breeding, multiplication, certification and distribution of quality seeds have always focused on the so-called food security crops to the disadvantage of high-value crops. For example, in Ghana tomato seeds are imported from France, Holland, Italy and the Czech Republic, among others or obtained locally through farmers' washed seed. Input suppliers such as Dizengoff Ghana provide technical assistance to client farmers through their consultants in the field and demonstration plots serving as training and testing sites for new seed varieties. Also, East-West Seed International has demonstration plots across the Ghana where they showcase its new seeds and train farmers on better methods of vegetable production. The company also provides introductory trainings for their customers and suggested plant protection plans (such as application of fertilizers and pesticides).

Testing and registration of new varieties on local seeds markets are limited by cumbersome regulatory procedures, which further constraints the competitiveness of local producers. Some producers are buying seeds from unofficial suppliers with some farmers producing their own seeds for their own use and for selling in small quantities to other farmers. Most local varieties however, are not appropriate for commercial production though they are still used by many open-field producers.

4.2.2 Seedlings Producers

There are very few known formal seedling producers that have the capacity to use advanced technologies and equipment for seedling production and to sign delivery contracts with farmers well in advance of the growing period. In many cases, individual farmers have to raise seedlings for themselves and for sale to their colleague farmers.

4.2.3 Fertilizers and Chemicals Dealers

There are many licensed companies and distributors of fertilizers and chemicals in Africa. Unfortunately, almost all fertilizers and chemicals are imported from outside the continent. It is only a few countries that produce fertilizers and chemicals locally. Although many small producers in most countries in Africa are less commercially oriented, they see fertilizer and chemical use as the only way to increase their output levels because they do not have great faith in the soils of their farmlands. In addition, there is lack of information on the proper application and storage of fertilizers and chemicals among producers. Almost all smallholder producers do not perform soil testing before the production season and apply fertilizers and chemicals without any fertilization and chemical application plan.

4.2.4 Packaging Suppliers

There are a number of packaging companies operating across commodity supply and value chains in Africa. However, the packaging materials and additives are often imported from Europe, Asia (China) and others.

4.2.5 Agricultural Equipment Suppliers

There are a number of agricultural equipment suppliers that offer various farm equipment such as tractors, cutlasses, sprayers, greenhouses and so on to farmers. Some of the equipment suppliers offer demonstration trainings and have websites with information about different kinds of equipment such as irrigation systems, greenhouses and ventilation equipment. They further provide some technical assistance to farmers related to different types of equipment and set-ups. Currently, many equipment companies are manufacturing equipment by order only and they do not keep stocks of such equipment.

4.2.6 Extension Services and Training

There should be readily available extension services to producers from both private and public providers, providing critical market information and production training to the entire commodity supply and value chains. The provisions of public extension services through agricultural extension agents (AEAs) to farmers are mostly by ministries responsible for agriculture but there are critical issues of relevance, effectiveness, efficiency, impact, sustainability and coherent with these arrangements in recent times because of poor funding by governments. Thus, many farmers have no access to extension services.

4.3 Regulatory Institutions

4.3.1 Quality Assurance Institutions

The structures of institutional support for the high—value agriculture sector in Africa are developing, albeit slowly. Institutions for monitoring the quality and safety of food products in the various countries such as Ghana include the Food and Drugs Authority (FDA) and the Ghana Standards Authority (GSA). The GSA sets the standards in relation to the quality and safety and monitors conformity to the set standards by individuals and companies. The FDA on the other hand ensures that all actors comply with the standards set by GSA. Thus, the role of GSA is conformity and that of GFA is compliance. These institutions must be proactive to ensure that farmers produce to the requirements of buyers, especially those in the high-value end markets such as supermarkets.

4.3.2 Agricultural Ministries

The ministries of agriculture have legislative and regulatory roles in the agriculture sector. For instance, in Ghana the Plant Protection and Regulatory Services (PPRS) Directorate of the Ministry of Food and Agriculture is responsible for regulating issues of crop breeding certification among others. The country passed the Breeders' Bill, which will incentivize breeders to engage in breeding plants and animals that respond to the needs of farmers and consumers. Thus, the ministry provides technical support and advice on range of issues including irrigation, agronomy, application of fertilizers, seedling raising, land preparation, use of certified seeds and regularly organize training for model farmers on modern agricultural production technologies, improved varieties, compost-making and demonstrations, among others.

4.4 Limited/Missing Services

4.4.1 Post-harvest Handling Facilities

The function of the post-harvest handlers in African agricultural commodity value chains is missing. No warehousing and pre-cooling/cold storage facilities or services are provided along the commodity value chains. Thus there is the need to build capacity in post-harvest handling so as to reduce post-harvest losses. This requires investments in pre-cooling vans and warehouses across the different clusters.

4.4.2 *Soil Testing Facilities*

There is limited soil testing facilities in African agriculture. The few soil testing facilities in many African countries are located distally from the farming areas where they would more likely be used. As a result, farmers are not performing soil testing. This leads to lack of information about the components of soil nutrients on specific farms leading to the unbalanced fertilization of farms, which brings about poor quality of produce. Currently, farmers consider the available soil testing expensive. Thus they are unable to test for Nitrogen, Phosphorus, Potassium (NPK), humus and pH, which should inform the type and quantity of fertilizer that should be used.

4.4.3 *Specialized Extension Agents*

Specialized extension agents with hands-on experience on high-value agricultural commodities production are in limited supply in the content. There are many of the production clusters where farmers do not have access to trainings that are tailored towards their specific needs. Farmers need to be trained by specialized extension agents on how to use modern production technologies and high-quality seed varieties among others.

4.4.4 *Financial Services*

Many farmers are using their own finances for production with only a few having access to bank loans. The low number of loans issued to farmers by banks is explained by high risks associated with agriculture in Africa with high probability of non-repayment for the banks. An initiative whereby farmers' investments in infrastructure including greenhouses and drip irrigation equipment could be reimbursed up to 50% of the investments needs setup by governments and its development partners is needed.

CHAPTER 5

Systemic Constraints Within Commodity Value Chains

Abstract Commodity value chains in Africa remain weak and fragile because of systemic constraints—stretching from production at the farm level through to output marketing. At the production level, majority of farmers continue to complain of inadequate access to financial services, improved planting materials, poor irrigation facilities, poor market infrastructure, high production and transaction costs, high prevalence of pests and diseases and poor access to knowledge and extension services, among others. At the level of the output market levels, the key constrains that make them sluggish include monopoly by 'market queens', lack of modern storage facilities for perishables and mistrust among others. There are also constraints that make it difficult for businesses along commodity value chains to strive.

Keywords Production constraints · Inputs constraints · Services constraints · Market constraints · Value chains

5.1 Introduction

Generally, farmers in Africa face numerous constraints in their production and marketing of farm produce. Some of the constraints include recurrent droughts and occurrences of pests and diseases that destroy

A.-R. Alhassan and M. Abunga Akudugu,
Supply Chain Management in African Agriculture,
https://doi.org/10.1007/978-3-030-54209-2_5

crops and/or reduce production. Famers are forced to sell their produce in distress because of the perishable nature of most farm produce exacerbated by inadequate storage and processing facilities resulting in low returns to investments and efforts. In other words, farmers are mostly not able to secure good value for their produce and this negatively affects their performance. There is limited shelf life as a result of poor post-harvest management. Post-harvest handling facilities such as pre-cooling, cold storage and refrigerated transport are almost non-existing in most of the countries. This calls for planting of the right varieties that would help extend the product shelf life.

Poor visual uniformity of products is also a challenge. Farmers mostly mix different sizes of their commodities as a result of lack of grading lines and calibration. Basic sorting and grading are done manually (if at all) and this affects the performance of the supply and commodity value chains. There is high incidence of visible defects due to unbalanced fertilization and pesticide use, which results in poor quality. There is a problem of compliance with food safety standards as a result of improper use (doses, timing, safety) of pesticides, insecticides, fertilizers and stimulators. This is exacerbated by the lack of recordkeeping on timing of sprays, so that the after-spray period does not overlap with the market delivery day leaving chemical residuals on the produce.

Failure to market cooperatively and recognize market standards among smallholder farmers leads to poor quality and lack of uniformity of produce supplied to end markets. This is also affected by inability to consistently supply sufficient volumes to high-value end markets such as supermarkets worsened by low productivity and seasonality. There is limited access to finance and lack of affordable credit resulting in low investments in technologies that would increase production such as cooling equipment for pre-cooling and cold storage and greenhouses. Fluctuating commodity prices seriously affect farmers' capacity for loan repayment. There is lack of planting and harvest planning, which results in farmers planting at the same time and therefore harvesting at the same time that bring about gluts and low market prices. There is also an issue of inability of farmers to harvest for an extended period of time in order to supply required volumes on a consistent basis due to poor timing of planting times. Besides, there is the need to plant new varieties that would support longer harvesting seasons. There is lack of long-term strategic planning and this limits farmers' ability to buy into new business models and cooperative marketing arrangements that would allow for collective access to higher value markets.

5.2 Production Constraints

5.2.1 Farm Finance

Across Africa, most farmers indicate that financing farm operations are a major constraint to production. Most farmers are resource-poor and find it difficult to finance their farming operations from their personal resources alone. Farmers are unable to obtain credit from financial institutions mainly because of lack of the requisite collateral security to buttress credit application. Most financial institutions are reluctant to lend to farmers as a result of the inherent risks associated with the sector.

5.2.2 Seeds for Planting

Many farmers have no access to locally developed improved seeds for planting. They, therefore, have to rely on so-called improved seeds imported from Europe, Asia and America as well as own selections from previous crops, and from exchanges between farmers. Most of these seeds are of low viability, as they are not selected by testing.

5.2.3 Irrigation Facilities for Dry Season Crop Production

Dry season crop production is a problem for most farmers across all regions and countries of Africa because of lack of well-developed irrigation facilities with the exception of north Africa. Only a small proportion of farmers produce through irrigation compared to the rain-fed farmers. Most part of countries in Africa experience mono-modal rainfall patterns with rains starting from May/June and ending in September/October each year. Those areas that experience bi-modal rainfall patterns have major cropping season, which begins in, mid-March to September and the minor cropping season, which is October to December. The period ranging from January to April/May is classified as the dry season, and rainfall during this period is generally low. Cropping in the dry season, therefore, implies that water must be supplied through irrigation. Occasional dry spells even during the rainy season, also does occur, creating the need for irrigation. Most farmers across the different regions of Africa do not have access to irrigation facilities, and even if they do, the cost of irrigation is considerably high. Those closer to natural water bodies such as rivers farm during this period using water pumps with an associated high cost of production. This problem, therefore, prevents most farmers from farming during this period thereby affecting all year-round supply food.

5.2.4 *Marketing of Produce*

Farmers always lament that low prices offered for their produce particularly during the peak harvest season constrain their farm enterprises. Across all countries, most farmers complain of the non-existence of guaranteed market and effective pricing mechanisms for their produce, which is a major disincentive to production.

5.2.5 *Cost of Production*

Farmers are often of the view that the cost of production is too high. This results in high cost of inputs such as fertilizer, agrochemicals (fungicides and insecticides), labour and cost of irrigation, if available. This situation farmers often note is worsened by lack of access to production credit exacerbated by high incidence of poverty. Thus most farmers are resource-poor and are unable to save enough to finance their farm productions.

5.2.6 *Land Preparation*

Farmers lament about the unavailability of tractors to plough fields at the right time as a constraint to their farm production. The unavailability of tractors to mechanize land preparation creates increased need and competition for human labour for land clearing/preparation and consequently, increases labour cost.

5.2.7 *Diseases, Pests and Weed Control Problems With*

The high incidence of pests and diseases is a major setback for farm production across the different countries. This is particularly so for those farmers engaged in rain-fed crop production, as environmental and climatic conditions are conducive for the occurrence of pests, diseases and weeds at this time. Farmers believed that these would not have been major concerns if they had access to improved and resistant crop varieties and animal breeds better suited for the local environments.

5.2.8 *Land Tenure or Land Acquisition Problems*

Some farmers, especially women and the youth have difficulties acquiring lands suitable for agricultural production. This is particularly so for farmers in the urban areas, especially migrant farmers who in some cases have to pay various sums of money for land rental. The acquisition of land presents a problem also for farmers who crop during the dry season. Most of these farmers usually prefer land close to a perennial source of water such as an irrigation dam or rivers to permit the use of their irrigation pumps, if they have or to embark on manual watering by using watering cans and buckets. Unfortunately, such lands are limited in supply in most production areas and the consequence is continuous cropping leading to pests and diseases build-up, which is detrimental to crop production.

5.2.9 *Seasonality*

There is a huge variation in crop productivity in different seasons. For instance a hectare of land under rain-fed produces less than same area of land put under irrigated production. The differences are due to reasons such as high incidences of pests and diseases in the raining season compared to the dry season. Farmer sensitization on correct techniques for different seasons is missing.

5.3 Inputs and Services Supply Constraints

5.3.1 *Access to and Availability of Good Quality Inputs*

Although farmers have ready access to inputs all year-round these inputs are often sold at high and unstable prices. There is also lack of access to quality and improved crop seeds that makes farmers to resort to using washed seeds. The input suppliers need credit for buying inputs in bulk and keeping a good volume in stock.

5.3.2 *Quality of Inputs*

There is huge trust deficit between input suppliers and farmers. Both point fingers at each other for the lack of trust. Issues of farmers using some fertilizers and pesticides that never work are commonplace. Whereas farmers blame some inputs suppliers for supplying poor quality and fake inputs to them, the inputs suppliers blame farmers for either not using the right quantities or not buying the right inputs for the right activities/times. The low literacy levels among farmers and input dealers, especially at the retail end exacerbate the situation. Most farmers and retailers do not have formal education. Most inputs retailers have not been trained in handling and utilization of agricultural inputs and are therefore not able to guide farmers on the proper uses of these inputs.

5.3.3 *Knowledge and Skills on the Application of Inputs*

Both input suppliers and farmers do not have adequate training on improved package of practices/right application of seeds, chemicals and fertilizers. Almost all of the actors agree that demand for fresh farm produce is steadily growing in local and national markets but the misuse of inputs by producers presents a serious challenge. Instances of farmers harvesting soon after the application of pesticides have been reported. Besides, some farmers engage in mixing different chemicals and applying to crops without the needed guidance by technical officers from agricultural ministries. This presents a critical health issue as it poses serious health risks to consumers.

5.3.4 *Availability of Credit*

The agricultural commodity value chain actors, especially farmers need cash to buy inputs and to invest in the purchase of production equipment. In fact, the cost of production using greenhouse technology is way too high as it runs into hundreds of thousands of dollars, definitely way beyond the imagination of majority of African farmers. Unfortunately, most actors are not able to raise this amount from savings and do not have access to credit to be able to effectively finance their activities. This is a serious constraint to the effective operation of commodity value chains in most of Africa.

5.3.5 *Poor Agricultural Extension Services*

Agricultural extension services are very important in enhancing productivity as it positively influences the adoption of modern production technologies. This is particularly so among smallholder farmers, majority of who have not had any formal education and therefore rely on extension agents for information on improved methods of production. Unfortunately, most farmers have no access to extension services due to inadequate public funding causes the situation. This is particularly so for horticultural crops as the focus of many African governments has long been on the so-called food security priority crops leaving the vegetable sector with no any serious interventions. Many Departments of Agriculture in many countries are poorly resourced and with most extension agents not having means of transport to visit farmers and deliver extension services. Where motorbikes are provided for this purpose, fuel is not provided at the right time for the extension agents to visit farmers. Besides, most agricultural extension agents are not specialized information in specific commodity production.

5.3.6 *Weak Research and Development Linkages*

There is weak support for research and development across Africa. Most research institutions have low interest in funding research by government and development partners. This explains why there is inadequate improved crop varieties specifically developed for different agroecological zones of Africa. This negatively affects the performance of agricultural commodity value chains because R&D is key for the advancement of any sector.

5.4 Output Marketing Constraints

5.4.1 *Dominance by the Network of Traders ('Market Queens')*

The marketing of vegetables is very much controlled by a network of traders in the destination markets. There is an informal cartel of traders popularly called 'Market Queens' that control activities in agricultural commodity markets and operate even across borders. They have very strong links vertically and horizontally. Horizontally they work in total coordination with each other. Most of them have been in this business for many years and have acquired some power by virtue of their experience of the markets to strongly influence agricultural commodity supplies and value chains. Indeed, they control entry into markets to the extent that producers are unable to sell their produce directly to consumers in such markets without passing through them. To a very large extent, they determine both the farm gate price and prices at the consumer markets. They do this by preventing producers from selling at the markets they control thereby creating gluts at the farm gates resulting in lower prices and creating shortages at the consumer markets resulting in higher prices.

5.4.2 Lack of Storage, Transport and Processing Facilities

There is lack of appropriate storage facilities for agricultural produce across the different countries of Africa. Pre-cooling and cold storage facilities in the production centres are absent. There are no temperature-controlled cargo trucks for the transport of fresh vegetables for example from production to consumption centres. In many African countries, large-scale processing facilities established in strategic locations to help reduce post-harvest losses through processing are currently not operating. The absence of these facilities leads to distressed sale of produce because of perishability and this is not good for the growth of the agricultural sector.

5.4.3 Fluctuating Market Prices

The market prices are generally guided by the trends in the big consumption centres in Techiman, Tamale, Kumasi, Accra, Koforidua, Cape Coast and Takoradi, among others. High prices in these big markets translate positively to high farm gate prices and the reverse is true. The prices for tomatoes and vegetables generally fluctuate quite frequently due to various factors such as religious celebrations (e.g. Christmas, Eidul-Fitr or Eidul-Adha). Overproduction/supply or otherwise during certain periods in the year among others also make prices to fluctuate. These fluctuations make investment planning difficult thereby affecting the performance of the tomato value chain negatively.

5.4.4 *Lack of Trust Between Traders and Farmers*

Many agricultural commodity traders have no positive experiences buying directly from farmers. Even after agreeing on deals, the farmers switch buyers even with marginal change in prices. Farmers on the other hand mostly accuse traders of always cheating them by offering extremely low prices for their produce. This makes long-term relationships with difficult thereby negatively affecting the viability of most agricultural commodity supply and value chains in Africa. It is important to build long and trusting relationships between producers and buyers in order to stimulate growth in commodity value chains, as traders have the capacity to leverage funding for the production activities of producers.

5.4.5 *Lack of Collective Bargaining and Marketing*

Although some farmers belong to social groups and producer associations, they never undertake collective marketing of their produce. Even though the collective members interact with same actors, they do not collaborate among themselves to bargain a better deal for their produce or even for inputs. This undoubtedly affects the performance of many commodity value chains across the continent, as the production segments of the chains remain fragmented.

5.4.6 *Poor Road Networks*

There is poor road network connecting production and consumption centres. Roads leading to some of the high production centres are impassible during the rainy seasons, which happen to be the period of major harvest. This phenomenon continues to dampen growth in the agricultural sector, as producers in some cases see their produce rotting away but are unable to salvage the situation because of impassible roads.

5.5 Business Enabling Environment (BEE) Constraints

Business enabling environment (BEE) constraints result in high costs of performing activities across agricultural commodity supplies and value chains in Africa. To truly transform the agricultural sector in Africa for sustainable and inclusive continental development, these BEE constraints must be addressed. Some of these constraints are outlined in the following Table 5.1.

Table 5.1 Business enabling environment under 5.5

Business Enabling Environment constraint	*Implications for the commodity value chains*	*Root of the problem/what needs to change*
Inadequate access to high yielding and disease/pest resistant crop varieties and animal breeds	Low productivity and yields affecting returns on investments	Low investments in breeding initiatives; Cumbersome procedures for testing and registering new seed varieties and breeds
Rising and unstable fuel prices	High cost of doing business for producers and marketers	Importation of petroleum products and instability of national currencies
Africa imports almost all of the fertilizers and chemicals used in agriculture	Low productivity, high cost of inputs	Lack of investments in manufacturing
Low quality of seeds and fertilizers offered on the African market	Low productivity, large amount of low quality products on the market	Need to increase the capacities of inputs testing laboratories
Difficulty meeting International Standards and Norms due to poor enforcements of quality standards compliant with global market requirement	Low ability to compete with imported agricultural commodties and to export to international markets	Need to reform legal, regulatory, and institutional framework for food safety and phytosanitary standards
Poor laboratory capacity for chemical testing, which includes pesticide residues, heavy metals, and mycotoxins (aflatoxin)	Consumer health and market access issues in higher value end markets	Need to build capacity for products testing and certifications
Poor logistics management (i.e. storage, transport and packaging)	High incidence of post-harvest losses and poor quality of produce at end markets	Need to invest in the provision of pre-cooling facilities and appropriate transport trucks
Poor information flows across supply and value chains	Poor performance of supply and value chains due to lack of feedback loops through information sharing	There is need for strong forward and backward market information sharing among actors across supply and value chains

CHAPTER 6

Benchmarking and Recommended Production Practices

Abstract In building robust and resilient commodity value chains, there must be strategic focus on the type of production practices that are adopted by producers. To be able to determine if progress is being made in a particular commodity value chain, there must be a mechanism for comparing the performance of the chain in question with the industry standard. In this chapter, we present how to benchmark the performance of a chosen commodity value chain with the best performing commodity value chain the industry standards. This is done using benchmarking and gap assessment principles. We also present the farm management practices required for building robust and resilient commodity value chains in Africa for sustainable and inclusive development.

Keywords Benchmarking · Farm management · Production practices · Gap assessment · Value chains

6.1 Benchmarking and Gap Assessments of Commodity Value Chains

The key performance indicators of most agricultural commodity value chains are reflected in the functioning of the actors in the inputs arena, services arena, production arena and output marketing arena. There are

A.-R. Alhassan and M. Abunga Akudugu,
Supply Chain Management in African Agriculture,
https://doi.org/10.1007/978-3-030-54209-2_6

no strong linkages among actors in the inputs, services, production and marketing arenas. Commodity value chains in Africa are benchmarked against the performance of those of continents. However, it is important to develop intracontinent benchmarking standards. For example, the tomato industry in Ghana could be benchmarked with that of other West African countries such as Benin, Burkina Faso, Cote d'Ivoire and Togo. Besides, intracountry benchmarking of different agricultural commodity value chains is also very critical and in our opinion most recommended. In this case, the seemingly best performing commodity value chain in the country is used as a benchmark to other commodity value chains in the country. For example, we could benchmark different commodity value chains to the pineapple value chain in Ghana. A well performing commodity value chain should have:

1. An efficient inputs arena that provides quality inputs at the right time, right quantities and affordable prices to farmers at the right places for production. A well organized and coordinated arena that is able to meet the needs and demands of other actors with free information sharing and collective investments decision-making processes for the benefit of all. Actors that are strategically positioned with the appropriate vertical and horizontal linkages with other actors for overall efficient performance of the value chain.
2. A services arena that effectively delivers the right services at the right places and time to actors engaged in carrying out various activities along the value chain. An arena that is well organized and coordinated in delivering the relevant services using state of the art methods and techniques through the appropriate vertical and horizontal linkages with other actors for overall efficient performance of the value chain. An efficient arena that is strategically positioned to provide services, which meet the needs and demands of its clients.
3. A well-informed and highly specialized production arena that is able to source the right quality inputs and services at the right time and prices for production with less risks and uncertainties. An arena that is well resourced, organized and motivated to take the right investments decisions in a coordinated manner for overall productivity of the value chain to the benefit of all. An arena with the requisite capacity to support the production of a commodity with differentiated characteristics that meet the demands of customers in the

different market segments using relevant and state of the art production technologies. Actors with the appropriate vertical and horizontal linkages with other actors for overall efficient performance of the value chain.

4. A market system that is able to command, organize and coordinate the different market segments in an efficient and effective manner for overall chain productivity. A well-resourced arena with the capacity to a commodity that meets the demands of consumers in the different market segments using the appropriate transportation, distribution, processing and delivery technologies driven by customer satisfaction. An arena with the appropriate vertical and horizontal linkages necessary for overall productivity of the value chain.

Given the industry benchmarks above, the key gaps in the performance of agricultural supply and commodity value chains in Africa are:

1. Inputs arena—poor quality seeds, expensive inputs, fake agrochemicals, agrochemicals overuse, poor knowledge on appropriate inputs use among farmers, poor vertical and horizontal linkages among actors in this and other arenas, etc.
2. Services arena—poor extension services delivery, poor access to financial services, poor access to agricultural advisory services, poor transport and storage facilities, poor vertical and horizontal linkages among actors in this and other arenas, etc.
3. Production arena—poor access to quality inputs including seeds, high input prices, poor access to mechanization and agricultural advisory services, inadequate irrigation facilities, poor rainfall patterns, poor soil fertility, low yields, low output prices, inadequate availability of production technologies such as greenhouses, poor horizontal and vertical linkages among actors in this and other arenas, seasonality, no comprehensive risks insurance, etc.
4. Output marketing arena—poorly developed market channels, no guarantee markets, high price volatilities due to seasonality, importation of cheap food products into the markets, poor transport and storage facilities, non-functional processing companies, poor vertical and horizontal linkages among actors in this and other arenas.

Table 6.1 Stakeholder benchmarking of the tomato value chain

Value chain drivers	*Pineapple value chain*	*Tomato value chain*	*Gap*
Input supplies	4	2	2
Services provision	4	2	2
Production processes	5	3	2
Marketing	5	2	3
Average chain function	4.5	2.25	2.75

Source Rating: 5 = excellent; 4 = good; 3 = average; 2 = below average; 1 = weak or poor

Using tomato and pineapple value chains in Ghana to illustrate the benchmarking process in Table 6.1.

The results indicate that the pineapple value chain has an average chain performance of 4.5, which is almost an excellent situation with the tomato value chain scoring an average chain function of 2.25, which means that the chain is performing below average. The driving force of the success of the pineapple value chain is targeted and concerted interventions in that industry by government, development partners (MiDA) and the private sector (Blue Skies). With similar attention and investments, the performance of the tomato value chain could be significantly improved. For instance, Blue Skies is able to within 48 h move pineapples from Akwapim Mountains, 100 km west of Accra, Ghana to final consumers in the United Kingdom (Blue Skies corporate Web site, www.bsholdings.com).

In terms of imports and exports levels, Ghana is a net importer of tomatoes and net exporter of pineapples (Table 6.2). The pineapple industry is obviously doing far better than the tomato sector. The fact that Ghana moved from no exports in the mid-1960s and early 1970s to the status of net exporter of pineapples from the mid-1970s to date means that the right steps were taken to turn things around. Thus, the tomato sector also needs similar pragmatic steps towards moving the country from a net importer of tomatoes and tomato products to a net exporter. Critical investments in improving the performance of the tomato value chain are of utmost importance.

Table 6.2 Benchmarking of the tomato value chain with pineapple value chain

Years	*Tomatoes (1000mt)*		*Pineapple (1000mt)*		*Difference*	
	Imports	*Exports*	*Imports*	*Exports*	*Imports*	*Exports*
1965–1974	12,701	0	0	0	12,701	0
1975–1984	3419	0	0	1400	3419	−1400
1985–1994	17,540	0	0	80,527	17,540	−80,527
1995–2004	165,188	17,789	22	304,857	165,166	−287,068
2005–2011	475,609	21,952	100	173,042	475,509	−151,090
Total	674,457	39,741	122	559,826	674,335	−520,085

Source FAOSTATS, 2015

Table 6.3 Value of tomato imports and exports: 1961–2012 ($1000)

Category	*Statistic*	*Benin*	*Burkina Faso*	*Côte d'Ivoire*	*Ghana*	*Togo*
Imports (I)	Minimum	250	0	12	4	0
	Maximum	17,023	8146	22,865	92,981	44,708
	Mean	4828	978	5681	12,815	2234
	Std. dev.	4202	1704	5139	23,734	6295
	Total	251,036	50,842	295,429	666,364	116,178
Exports (X)	Minimum	0	0	0	0	0
	Maximum	2649	3143	242	12,556	1391
	Mean	113	279	48	500	64
	Std. dev.	388	631	57	1926	212
	Total	5867	14,507	2485	25,986	3350
Trade deficit (I-X)		245,169	36,335	292,944	640,378	112,828
Trade burden (I/X)		43	4	119	26	35

Source FAOSTATS, 2015

Using intercountry benchmarking, a comparison of the levels of imports and exports of tomatoes in Ghana with her neighbours indicates that Ghana is not performing well given its vast potential for tomato production. For instance, between 1961 and 2012, Ghana spent US$666 million on tomato imports compared with US$251 million for Benin, US$50 million for Burkina Faso, US$295 million for Côte d'Ivoire and US$116 million for Togo (Table 6.3).

Although Ghana did not do badly in terms of tomato exports during the period under review (i.e. 1961–2012), it needs to do more as the country managed to get about US$26 million from tomato exports over the period compared to the about US$6 million for Benin, US$15 million for Burkina Faso, US$2.5 million for Côte d'Ivoire and US$3.4 million for Togo (Table 6.3). In terms of trade burden defined as the ratio of exports to imports, Burkina Faso is the best performing country among its peers importing just about 4 times its exports compared to Ghana importing about 26 times of her exports, Togo doing 35 times of her exports, Benin 43 times and the worse performing is Côte d'Ivoire with about 119 times of her exports.

6.2 Farm Management Practices

6.2.1 Suitable Varieties

For improved productivity, farmers must make use of suitable crop varieties that should target at meeting the taste and needs of consumers or the target markets. For example, farmers who target the processing channel must cultivate varieties that are preferred by processors.

6.2.2 Source of Planting Material

Most farmers are currently not using certified crop seeds. There must be intensified efforts to ensure that farmers buy certified seeds from reputable seed companies.

6.2.3 Climatic Requirements/Site Selection

High temperatures and low humidity lead to reduce yields drastically. For better results therefore soils should be well-drained and fertile.

6.2.4 Land Preparation

A thorough land preparation is important in enhancing early crop establishment and adequate weed control. Farmers are recommended to prepare ridges or beds across contours on which seedlings or seeds may be transplanted or sowed and to construct a farm pond to collect excess water and reuse for irrigation.

6.2.5 Propagation

Seeds can be sown directly or nursed for transplanting. Depending on the type of crop, seedbeds should be 1.2 m wide and at any convenient length and then level beds. Water beds, cover with dry grass and burn or solarize soil with transparent plastic sheets for 5–8 weeks to sterilize the soil. Sow seeds in drills 10 cm apart. Cover beds with well-dried non-seeded grass or palm fronds. After emergence removes dry grass and provides shade over the bed. Thin out weak, malformed seedlings to avoid overcrowding. Prick out seedlings at first true leaf stage. Transplant seedlings 3–4 weeks after emergence on the field at 5-leaf stage. Two weeks before planting on the field, apply a liquid feed of 5 g/L of NPK 15-15-15. Harden seedlings 1 week before transplanting by decreasing shade until at least 1-day full exposure to sunlight and/or reducing irrigation.

6.2.6 Planting/Sowing

Planting or sowing should be done preferably late in the afternoon. The recommended planting distances for many crops are 60 × 30 cm in the dry season and 60 × 60 cm in the wet season. Seed-planting rate ranges from 300 g/ha to 450 g/ha.

6.2.7 Weed Control

Timely weed control is necessary for healthy crop growth. This may be achieved by frequent shallow hoeing. Application of pre-emergence weedicides 3–4 days before transplanting or sowing will enhance weed control.

6.2.8 Irrigation

Water supply is very important, especially in the dry season. The most critical time for ample soil moisture is during bloom and early fruiting stages.

6.2.9 Nutrient Requirements and Fertilizer Application

For better yields, farmers must make sure the soils contain the appropriate quality and quantity of nutrients specific to the crops grown.

Varying combinations of N, P_2O_5, K_2O, Ca and S, among others may be required. Use soil test as a guide to fertilizer application. For example, application of commercial foliar sprays with adequate amount of Boron and Manganese would enhance tomato quality and yield.

6.3 Pests and Diseases Control

Major crop's pests are aphids, grasshoppers, whiteflies, crickets, leaf miners, beetles, mites and caterpillars. These could be controlled with the application of potassic soap solution, insecticidal soaps or recommended insecticides. For instance, the major diseases affecting tomatoes at pre-planting and later stages include:

a. **Damping off**—This disease attacks seedlings on nursery beds and could be controlled by selecting well-drained areas for the nursery and regulating irrigation. Solarize the soil for 4–6 weeks before planting, or apply recommended fungicides e.g. copper-based fungicides.
b. **Early blight**—This disease attacks seedlings at the seedbed stage and high temperatures may favour disease incidence. Using disease-free seeds and pricking out into individual containers could control it.
c. **Septoria leaf spot**—The disease attacks the plant at any stage. Under heavy infection, spots occur on the stems and flowers. Burning all tomato remains after harvest, practising crop rotation with cereals and legumes or use of fungicides could control the disease.
d. **Sclerotium wilt**—Attacks the base of the tomato stems. Affected plants show drooping leaves and finally a complete wilt. Control by crop rotation, good field sanitation and by the use of recommended fungicides.
e. **Root-knot nematode**—Nematodes feed on young root causing wilt and this leads to reduction in fruit yields. Applying recommended nematicides, use of tolerant varieties and crop rotation using cereals and application of well-decomposed poultry manure can control it.

6.4 Yield, Harvesting and Market Requirements

Yield varies greatly with crop variety and adherence to good agricultural practices. Every crop has potential and achievable yield rates depending on the varieties and agronomic practices. Harvesting must be done at the right time in order to avoid losses and also to meet market requirements. Thus, the right time to harvest to a large extend depends on market demands and distance.

CHAPTER 7

Market Opportunities and Upgrading Needs

Abstract Agriculture in Africa is the main source of employment to many but largely not modernized. This means there is urgent need to upgrade the key structures for success. In this chapter we argue that building robust and resilient commodity value chains for equity, inclusive and sustainable development requires improvements in strategic and operational productivity. It is also important to improve the efficiency in managing supply and commodity value chains as well as improving the human resource capacity for efficient supply chain management. This means upgrading and deepening through adding new operations, specialization and new investments, among others. We also argue in the chapter the need to create the enabling business environment taking into consideration cross-cutting issues like such as gender.

Keywords Market opportunities · Upgrading needs · Operational productivity · Strategic productivity · Enabling environment

A.-R. Alhassan and M. Abunga Akudugu,
Supply Chain Management in African Agriculture,
https://doi.org/10.1007/978-3-030-54209-2_7

7.1 Introduction

Upgrading needs of agricultural commodity supply and value chains are determined by market opportunities and overall market positioning strategy. The following market opportunities have been identified for African producers, investors and agri-preneurs:

1. Domestic market import substitution focusing on the local supermarkets and processing channels by extending the production season.
2. Building reputation and reliable long-term relationships with high-value end markets, particularly supermarkets by increasing the quality and consistency of locally produced agricultural commodities.
3. Building reliable relationships with domestic agricultural commodity companies in order to have diversified channels.
4. Upgrading opportunities in the entire value chain driven by the requirements of targeted market segments in end markets and addressing systemic constraints within the value chains that prevent the sector from meeting the critical requirements of these markets.

7.2 Critical Success Factors, Constraints and Segment Deficiencies

The analysis below is focused on critical success factors in each of the targeted market segments, improvements required and constraints that need to be addressed (Table 7.1).

Generally, some deficiencies must be addressed so as to promote efficiency and productivity in agricultural commodity value chains in Africa. These deficiencies and possible solutions are broadly summarized in the following Table 7.2.

7.3 Upgrading Needs

7.3.1 Improving Strategic Productivity

To improve the productivity of agricultural commodity supply and value chains it is important to choose the appropriate markets and products by adding new product features and service components that enable actors to

Table 7.1 Table on critical Success factors, constraints and segment deficiencies under 7.2

Market segment	*Critical success factors/market requirements*	*Improvement needed*	*Constraints to achieving these improvements*
Supermarkets	Long shelf life (at least 1 week) from the delivery time	Choosing the right hybrid and timing of planting & harvesting; planting varieties with high firmness; pre-cooling facilities	Expensive hybrids, lack of knowledge about appropriate varieties and lack of pre-cooling facilities
	Supplier capacity to offer supply over extended period	Cultivating hybrids with different ripening periods; planning of planting and harvest times; adoption of greenhouse and irrigation production systems.	Increase producer awareness that such business model is profitable; Lack of technical background for new varieties; High cost of greenhouse facilities
	Ensured delivery of contracted volumes of commodities	Intensive production technology using greenhouses, irrigation and hybrid varieties; Identify dependable suppliers providing volumes on sustained basis	Lack of Irrigation; High costs of inputs, fertilizers, planting materials; Farmers unwillingness to supply raw material when the open markets have higher prices
	Uniformity of product	Use of cold chain (from pre-cooling to truck delivering product); Use of appropriate hybrids; Use of calibration machines; Customized packaging	High investment cost for pre-cooling and packaging; Lack of knowledge about "extended shelf life" varieties; Lack of uniform quality from divided production communities; Lack of understanding among farmers about the required quality and quantity
	Seasonal contract	More frequent negotiation of prices in different supply periods that can better reflect the current market situation to limit risk	Price fluctuations on open markets; Low bargaining power of producers

(continued)

Table 7.1 (continued)

Market segment	*Critical success factors/market requirements*	*Improvement needed*	*Constraints to achieving these improvements*
	Invoiced transactions	Need for a central warehouse to enable pre-cooling and very short-term storage; Better production planning and harvest schedules	Comparatively high cost of cooling equipment; Lack of awareness among farmers about this business model; Limited production planning and harvest scheduling
	Ability to accept post-payment terms	Producers need to accept the terms where they are not paid in cash on the spot	Limited cash flow of producers; Lack of financial planning
	Supermarkets need to deal with a small number of suppliers	Need to identify a few suppliers with adequate supply capacity or joint marketing efforts of small producers	Inconsistency of produce in the given period
	Suppliers need to have Quality Certifications	Farmers need to obtain certificates for their production and to reinforce phytosanitary inspections	Lack of information about certification; Additional payment needed for certification; Lack of knowledge about keeping records on fertilizers and pesticides used; Limited local/national phytosanitary inspection and testing capacity
Processors	Ensured delivery of contracted volumes of commodities	Intensive production technology using greenhouses, irrigation and hybrid varieties; Identify dependable suppliers providing volumes on sustained basis	Lack of Irrigation; High costs of inputs, fertilizers, planting materials; Farmers unwillingness to supply raw material when the open markets have higher prices
	Guaranteed quality of supply (e.g. tomatoes for processing)	Following the technological requirements in crop production; Identify dependable suppliers providing required products needed for different specs	Lack of specific skills pertaining to commercial varieties; Lack of Irrigation; Lack of open/established relationships with producers who can honour the supply contracts as agreed

Market segment	*Critical success factors/market requirements*	*Improvement needed*	*Constraints to achieving these improvements*
	Reasonable cost for raw supply	Understanding the adequate pricing for raw supply; Proper cost analysis to determine % contracted versus open market	Lack of economic analysis and unwillingness to understand the benefits of the business model; High production costs; Unpredictability of the open market prices of tomatoes
Open air markets	Extended shelf life of product; Uniformity of product; Producer consolidation in production areas	Use of cold chain (from pre-cooling to truck delivering product); Use of appropriate hybrids; Use of calibration machines; Customized packaging	High investment cost for pre-cooling and packaging; Lack of knowledge about "extended shelf life" varieties; Lack of uniform quality from divided production communities; Lack of understanding among farmers about the required quality and quantity
	Availability of right varieties; Packaging; Ability to offer consistent delivery.	Detailed knowledge about supermarket requirements in different end markets; Use of required hybrids	Lack of knowledge about "extended shelf live" varieties; Lack of uniformity of delivered tomatoes; High /unaffordable cost of modern packaging; Attracting more customers to loyally buy Ghanaian produce which is tastier and brings incomes to the pockets of the local farmer

Table 7.2 Table on deficiencies and Possible solutions in 7.2

Segment	*Problems/Deficiencies*	*Requirements/Solutions*
Cultivation	Production faces problems such as diseases, nematodes, insect pests, high flower drop, low yield; Pesticides and herbicides can contaminate drinking water; Poor quality fruit; Needs good weather condition	Provision of resistant varieties; Sensitization of farmers on fertilizers and pesticides and their use; Cleaner Production; Providing finance for seeds, fertilizer and pesticides
Conversion	A lot of waste due to lack of processing and preservation facilities; Lack of transport/logistics; High cost of operation of processing machines that affect cost of final good; Obsolete processing machines with no spare parts; Interruption of supply when price of tomatoes in the open market rises as farmers will prefer to sell to markets with high prices	To set up multipurpose modern efficient processing factories equipped with state of the art machinery; Possible integration through pre-financing of seeds and fertilizer/pesticides by processing company to farmers, which will in turn guarantee a steady supply of products
Distribution	Poorly developed logistics sector; poor transport system; cheap imports often compete with local fresh and processed products	Developing robust local logistics sector; improving transport systems connecting production areas to consumption centres; Increasing the productivity of local producers to produce at the lowest per unit cost

increase the value of their products and prices. This "strategic" or 'market price-related' productivity is as important as 'operational' productivity in determining competitiveness. Efforts to strategically position commodity value chains in key agricultural markets must be intensified to orient and focus investment on the highest value activities. By positioning and/or repositioning fresh and processed commodities in new markets, to new customers or by adding additional value (real and perceived) are important in increasing sales and profits per unit of capital and labour. The product segments that are currently being offered, especially the fresh commodities are not assorted enough for the different market segments. The highest value chain segments are the fresh commodities for supermarkets and hotels for salad and processors, which are currently underserved. The open-air and roadside markets are the ones currently served.

7.3.2 *Improving Operational Productivity*

Improving productivity in commodity value chains requires increases in operational productivity. This is could be achieved through improvement in technology, manufacturing and service processes within specific segments of the value chain. Thus improving operational productivity in the commodity value chain requires that actors focus on improving the efficiency of each segment of the value chain through the implementation of the appropriate activities. Thus factor productivity must be increased at the farm and non-farm levels of the value chains using the appropriate technologies such as greenhouses and improved varieties. It is also important to introduce new technologies that improve processes and management systems. This will ensure that key players in the value chain can lower their costs and raise the productivity of their businesses, and the value chain overall. This is important if locally produced agricultural commodities are to be able to compete favourably with the imported ones. This will involve reorientation to ensure that Africa's agriculture is not excluded from competitively serving key markets because of the cost's structure. While ensuring that actors along commodity value chains are able to produce at the lowest costs possible, it is important to balance the trade-offs between cost and quality of crops produced for the various segments of the market. Thus the key opportunities for lowering costs without compromising quality must be sought at all times. This means that the actors and activities that are in control of the cost drivers must be constantly monitored to ensure continuous productivity. This requires that actors continue to employ analytical tools such as:

a. Per unit economic cost–driver analysis (fixed versus variable costs within each activity)
b. Cost–trend analysis (historical and projected changes in costs drivers)
c. Per unit activity–cost benchmarking.

7.3.3 *Improving the Quality of Supply Chain Management*

The tertiary way of improving productivity in commodity value chains is to improve the quality of management of the supply chain. Thus there is the need to focus on the effective management of costs of raw materials, transportation logistics, communications and information technology. Given the dynamic nature of commodity value chains, there is the need to continuously monitor the flow of goods and information along the chain of activities, the efficiencies of these flows, the transactions that facilitate value added, and the economic relationships that underlie each set of transactions. The relationship among buyers and suppliers along the value chain must be organized in such a way that it is cooperative rather than adversarial. It is important to look at the effective flow of information along the value chain (market trends, changes in price, external cost pressures) and the awareness level of producers of the downstream market dynamics of the industry (market trends, demand conditions, pricing). Efforts must be made to ensure the continuous availability of supporting services (financial, logistics, administrative) across activities in the value chain.

7.3.4 Improving Human Resources Across the Value Chain

Improvement in the value chain productivity requires availability of quality human resource base for the chain to tap. Thus improving productivity in the value chain requires investments in human resources. This involves enhancing motivation, management and training actors along the value chain. Improving the overall quality of the workforce along the value chain should be the mandate of the government and the educational institutions that are outside the direct control of the specific commodity industry. However, industry actors should handle the specific technical and management skills of actors along the value chain. Thus upgrading the technical and management skills of the entire value chain requires close cooperation of the actors along the different segments of the chain with support from government and academic institutions. There is the need to provide incentives to encourage actors along the value chain to invest in their technical and management skills development. In this regard, there should be supporting educational services (industry certifications, IT training, technology application) available to actors locally to increase their management skills levels. This also requires that academic institutions know the needs of industry actors and so align their curricula to meet industry needs.

7.3.5 *Improving the Business Environment*

The quality of the business environment ultimately serves a gatekeeping function, and often a negative one. Productivity increases require improving the quality of business strategy and operations by actors in the tomato value chain. For example, by forming new partnerships, actors are able to access markets, technology and finance, among others. There should be interventions by government and its development partners that will enhance productivity across the four components (strategic productivity, operational productivity, supply chain management and human resources) discussed above. However, a factor common across all value chains and components is that the business environment contributes importantly to the ability of actors to succeed in enhancing productivity. The following are the 'business environment' factors or the microeconomic foundations of growth that must be focused:

a. Level of bureaucracy and red tape;
b. Extent to which the rule of law is enforced and commercial courts are functional;
c. Quality of infrastructure;
d. Level of financial sector modernization and regulation;
e. Levels of trade access; and
f. Ability of the country to attract foreign investment and to ensure the proper functioning of land markets (through reliable registries) and labour markets (through policies that encourage job creation, labour flexibility and overall productivity).

7.4 Upgrading and Deepening Commodity Value Chains

Upgrading and deepening commodity value chains depend largely on the profitability, which is supported by services that allow the chain to grow, be more efficient, and enhance its competitiveness. The overall objective is to improve the depth and breadth of services currently being provided to the value chain, enabling them to be commercially sustainable and to help those services emerge where they are not being provided. The availability of new and better services should enhance the profitability of the whole value chain. Commercially sustainable services will typically involve specialization on the part of service providers who play critical roles in the deepening of the value chain. The range of services that can add value and strength to the value chain is vast. Examples include input supply, market information and product development support, business management and consulting, transportation and logistics, quality assurance (including certifications), skills, extensions and training and credit and other financial services.

Services can be delivered through many appropriate mechanisms. Services along the value chain can be provided by both public and private entities using commercially sustainable approaches. Additionally, services may be delivered as part of another commercial transaction, such as in the case of embedded services. For example, a processor extends credit in the form of 'virtual' working capital to a small-scale farmer when the processor provides seeds, fertilizer or pesticides as part of an out-grower scheme. In this context, for example, formal financial institutions can enter the equation and make credit flows available, perhaps even extending the offer to other financial services (savings, transfers and longer-term loans). This would extend the depth, breadth and sustainability of the services that were previously provided (mainly short-term working capital). This implies that the demand for services can also offer opportunities for small and medium enterprises to enter the value chain market as suppliers.

To upgrade and deepen agricultural commodity supply and value chains therefore, it is useful to map the particular services that are currently being provided (their sustainability, quality and location within the chain), as well as those services that are potentially viable that can improve the chain's performance. How to upgrade and deepen commodity value chains in Africa are briefly discussed in the following subsections.

7.4.1 Upgrading and Deepening Through Adding Operations

To upgrade and/or deepen commodity value chains in Africa, actions driven by market opportunities and demands must be undertaken. These actions should be guided by sound market analysis, strategic planning and good communication among value chain participants including input dealers, services providers, producers, traders, processors and consumers at the end market. The possible operations that could be added in order to upgrade and deepen specific commodity value chains in Africa are breeding of improved varieties for local environmental conditions, small-scale on-farm and off-farm processing equipment, improved storage and transportation facilities, among others. There is also the need to establish competitiveness initiative working groups at each segment of the value chains to work with chain participants to encourage innovation.

7.4.2 *Upgrading and Deepening Through Specialization*

There is the need for specialization by the value chain participants. For instance, input dealers need to specialize in providing inputs specifically needed for production of given crops. The opportunities for specialization should be guided by the size of the market for the specialized operation or service, and on the confidence that the customers of the specialized business or operation have that the work will be carried out to an appropriate quality level with needed degrees of customer service, requisite confidentiality and on a sustained basis. That is, the specialized business should be such that it does not fail and disappear. These criteria are, in many respects, the entry criteria for value chain deepening through specialized operations. Undoubtedly, upgrading and deepening the value chains through specialization will create the opportunity to accomplish otherwise unattainable goals. It will lead to risk and investment costs been shared and offset by the cost savings that result from cooperation and information sharing among actors.

7.4.3 *Upgrading and Deepening Through New Investments*

Attracting new investments from domestic and foreign sources are key in deepening and upgrading commodity value chains. Foreign direct investments by multinational corporations ensure that technologies are transferred into the local value chains. Also, knowledgeable domestic businesses should be encouraged to purchase or license new technologies in the local industries. Bringing quality sorting or product-packaging operations into commodity supply and value chains will increase cost efficiency and product competitiveness while adding value to the product value chain and local economy.

7.4.4 *Upgrading and Deepening Through Vertical Integration*

Vertical integration requires that actors in commodity supply and value chains continuously identify opportunities to incorporate new technologies or operations into the various segments of the chain in which they operate. Actors must recognize opportunities to add value, achieve efficiencies or improve quality. Adding operations through vertical integration enables value chain upgrading and deepening. In many parts of the world, vertical integration is achieved through consolidation. The formation of new organizations, such as farmers associations, service-provider associations and marketing organizations that will provide opportunities for otherwise fragmented producers and other businesses to combine their resources to add operations to the value chains.

7.4.5 *Upgrading and Deepening Through Joint Commercial Ventures*

Joint commercial ventures in the value chains will promote risks sharing among actors. This will lead to additions of new operations into the value chains or investing in new technologies that will enhance chain productivity. Produce preparation, packaging and logistics centres are frequently good candidates for joint ventures and should be explored. A joint venture among a cluster of actors including academic institutions for research and development (R&D) as well as value chain participants and entrepreneurs for investments is crucial in developing and commercializing African agricultural production, processing, storage and transportation technologies for overall chain productivity. This could be achieved through the establishment of commodity strategic working groups.

7.5 Business Model Identification and Replication

It is important to identify among the proposed business models (see Sect. 3.6) the appropriate business model for replication and this should be guided by relevance, practicality and what will encourage entrepreneurs to invest in and to manage the specific value chain for enhanced productivity. To finance the chain, entrepreneurs should put up a per cent of the investment and given a no-interest loan for the remainder to be subsidized by the government through designated financial institutions (DFIs). Agricultural ministries and other relevant bodies should provide technical training on how to operate the chains. Downstream actors (input suppliers, a section of service providers and farmers) should be committed to effectively play their roles to ensure that midstream actors (small traders, wholesalers and farm associations) get adequate quantities of produce for upstream actors (supermarkets, industrial processors and on-farm processors). Thus actors should be committed to ensure regular and predictable supply of inputs, services and products along the entire chain. To achieve this, three key areas must be of priority and these are briefly discussed in the following subsections.

7.5.1 *Communicating the Case for Investment*

It is important to effectively communicate the business case to entrepreneurs, financial institutions, political leaders and policymakers and implementers for them to buy into the idea. Thus the identified business models should be presented to potential entrepreneurs, financiers and other actors along the value chains for them to evaluate the market opportunity. Cluster level effective communication of the business models to actors, especially financial institutions that have the financial power to make things happen is crucial for the success of the business model.

7.5.2 *Product Positioning*

There is the need to properly position the identified products (fresh or processed) within the end markets. Thus producers must produce quality products that conform to tough international standards. This would encourage end markets to source commodities locally instead of relying on imports from other countries. Farmers should be encouraged to produce to meet the EUREPGAP and GLOBALGAP specifications. To leverage the quality improvements made within the value chain, clusters must strategically position themselves within local markets that have similar requirements as export markets by using branding, certification, unique shelf space and packaging that promote quality and freshness. Actors must make these choices with purpose so that they are not out-positioned by their competitors. This means that understanding the competition is a critical aspect of product positioning.

It is important to note that movements to new positions do not have to happen instantly but rather changes should be adopted incrementally, for example, by adding differentiation through higher quality or better variety, and then adding product scope through processing or improved packaging. This could be achieved through branding and certification; creation of unique shelf space and packaging. Branding and certification are important. The commodity value chain working groups should undertake some local advertising of their brands through the appropriate communication channels available in each cluster, region or country. The value chain working groups should lobby supermarkets and grocery stores to allocate dedicated shelf space for locally branded and certified products. A unique packaging that highlights product freshness and high quality should be created.

7.5.3 Increasing Product Value Through Standards

Choosing and targeting standards should be part of the strategic steps to transform commodity supply and value chains in Africa. Many export markets have standards and/or certifications that are needed for entry. By achieving international certification or standards, locally produced agricultural commodities have the opportunity to compete with imported ones thereby offering local value chain actors the chance to select the market positioning of their products. Also, value-added standards will allow for entry into certain niche markets such as supermarkets, which are patronized by the growing middle class and expatriates across Africa.

Implementing the processes and systems to meet standards requires resources and this means that it is important for value chains' leaders and firms to select the most strategically appropriate target market segments. Comparing implementation costs and local value chains' capacity to incorporate standards against the benefits of selling, the selection of high-value end markets (e.g. for supermarkets and processors) is a strategic choice. Thus implementing standards within a value chain involves three main elements and these are certification bodies, information channels and management as well as governance structures. The relevant certification bodies in Africa that have the capacity to certify that products meet certain standards must be strengthened to work. These are governmental bodies that are mandated to certify products and services to protect citizens' health and welfare. In addition to these bodies, governments should encourage the setting up of private certification institutions through public–private collaborations. This will ease the envisaged pressure on the governmental certification bodies.

Actors working in the commodity value chains must be motivated to change the way they produce goods and services to meet standards. This is only possible if the appropriate industry information and economic incentives are provided. These incentives must include shared knowledge of the requirements (and ideally, the logic behind them), as well as price points that reflect the additional costs and work involved in meeting requirements. There must also be trust within value chains that the process is stable. One aspect of creating trust is ensuring that proper information channels are available and being used. These will give value chain actors the confidence that they will obtain fair rewards for the costs of implementing new processes to meet standards. This means that the agricultural commodity value chains must communicate formally and informally. Formal communication can occur through meetings, cluster activities, integrated supply chain systems, industry associations, conferences and exchanging price information. Having a transparent culture within the value chains can create informal channels. For example, if a farmer sees a colleague benefitting from an investment to meet standards through better prices, he or she is more likely to upgrade processes based on the observation.

A value chain that is upgrading its standards requires some institutions to take responsibility for managing the supply base to meet standards and specifications. Because the commodity value chains are made up of independent actors with different motivations, this can be a difficult task. Therefore credible governance that clearly links value chain performance to market rewards is needed in order to be able to coordinate actors to work towards satisfying the needs of the target markets. In line with the proposed business models, three main types of institutions for management of the value chains in Africa are proposed and these are value chain management instituted by:

1. A lead actor (e.g. an input dealer, producer, off-taker, etc.)
2. Farmer-based organizations and marketing cooperatives
3. Government agencies.

The success of these various forms of value chain management and governance rests with their ability to understand market requirements and translate them into in-chain procurement standards, communicate information effectively, and motivate suppliers to respond with needed investments and operations. Lead actors should view the effective management of the value chain as an opportunity for growth creation and larger profits rather than simply as an additional imposed cost. This should be based on the fact that if the participants in the value chains work together, the value chains can more easily meet changing the markets needs and reduce transaction costs. Therefore, the lead actor should take responsibility for managing the chains by ensuring that each actor is meeting market standards.

Value chain management and governance system based on farmer-based organizations and marketing cooperatives can be very effective in clusters where there are fragmentations as producer organizations and marketing cooperatives can help manage value chain elements to meet market standards. These institutions are often important when first implementing standards in a value chain; institutions that are producer owned, visible and well known can build trust in the standardization process. Properly resourced, they can be very effective in training and outreach activities. It is important to involve credible third-party institutions (e.g. universities and research institutions) to help overcome initial mistrust between producers and processors/exporters. This will help focus the activities of all value chain actors on a coordinated approach to meeting standards.

Government agencies can play the role of industry governance, including standard setting, inspection, extension and communication. However, it is not recommended that state institutions lead chain management and governance because they are overburdened and poorly resourced that they might not be effective in their management and governance of the commodity value chains. Government agencies are slow to understand and respond to market trends thereby limiting their strategic focus and inflexible in promoting varying standards. This coupled with public sector inefficiencies, political and budget pressures will not allow government agencies to be able to effectively manage and govern activities in the agricultural commodity supply and value chains.

7.6 Creating the Enabling Environment

Creating the appropriate enabling environment for businesses operating along commodity value chains is critical. Constraints in the enabling environment can limit the development and growth of the chains through increasing costs, decreasing competitiveness, decreasing products and service quality, increasing businesses and investment risks and decreasing investments. Development of long-term strategies determines the ability and willingness of value chain actors to upgrade and make strategic choices. A poor business environment will definitely limit capital investments in the value chains. Reforming policies so that they are favourable to businesses can encourage dynamism and efficiency. The integrated nature of value chains makes them susceptible to constraints in the enabling environment at any level. Constraints affect their costs and ability to act, from the most basic supplier of raw materials to the processor.

Governments are often the principal actor in changing regulatory issues and are responsible for improvements and reforms. Often the private sector's interactions with governments (and those of development partners as well) tend to be on the national level. However, to a value chain, the capacities and wisdom of local government authorities are more important due to their impact on strategic portions of the value chains. Lacking incentives and experience, local government authorities tend to be less sophisticated in their understanding of the constraints faced by value chain participants in the enabling environment. Businesses within the value chains will have more personal relationships with local officials, which can be both a benefit and a barrier to enacting pro-growth change. The private sector must understand that effecting change is crucial for its survival and should therefore see promoting change its responsibility.

7.7 Public–Private Dialogue (PPD)

Through both official and ad hoc forums and vehicles for dialogue between the public and private sectors, agricultural commodity value chains can communicate their opportunities and challenges within the enabling environment. By sharing industry strategies that are developed through analytical tools such as positioning and benchmarking, the public sector can understand the impacts of its actions, decisions and capacities and more effectively prioritize issues and actions to align with private-sector needs. Public–private dialogue could be achieved by establishing mandates that include memoranda of understandings (MOUs) and formal government mandates or temporary activities. It could also be by alignment of the chains with key state institutions.

The alignment process might include where the headquarters or secretariat of a proposed Value Chain Working Group should be located. The proposed working group should have a clearly defined structure and rules for participation with the secretariat serving as the coordinating body. The working group members will come from various segments of the value chains and possibly donors. Participants will be decision makers and influential members of the value chains who want to facilitate change. As broad and representative a group as possible should be chosen and should include women from each segment to the extent possible. A strong but not overly powerful actor should be chosen from the chain participants to represent and drive the process by serving as a broker and helping participants negotiate and set timetables and priorities. This includes determining the types of outputs the PPD process will generate and could take the form of roadmaps, a policy white paper or reform proposals in addition to softer outputs such as cooperation, which emerges through dialogue. Quantifiable results early on can help build momentum and credibility for the process. In all these, public outreach and communication are important for disseminating broader-scale changes in the enabling environment and generating larger-scale buy-in for reforms. It can include branding of the PPD initiative, media education and awareness campaigns and social marketing.

The working groups should document clear successes and obstacles so as to build trust and generate traction for change. It is important to have channels of communication open with continental, regional, national and subnational government authorities because they each have different influence on limiting or facilitating investments and growth in the value chains. For example, processing companies cannot take advantage of tax holidays offered by government if the roads leading to the production areas are impassable. Thus actors in the value chains must know which levels of government determine the policies that most affect their operating environments and work to engage them. In order to effectively communicate as a unit, value chain participants must first develop a clear consensus strategy that outlines and prioritizes the policy changes needed. The value chain working groups' concept can be a tool for encouraging policy changes that enhance the competitive position of actors within the value chains and link the value chains' strategy to the dialogue process.

The working groups' success rests on several factors that affect the quality of the dialogue and its strength as a forum for advocacy. First, the private sector must clearly understand the public sector's role in the value chain development. The public sector may often be blamed for, or looked to for solutions to problems that are or should be more in private producers' control. It is also helpful to understand that the policy and regulatory environment have both legal and enforcement components and these may differ substantially from chain segment to segment. For instance, the policy and regulatory environment for input dealers might be different from that of traders or off-takers.

For long-term relevance of the working groups for public–private dialogue, regular benchmarking of specific business environment elements is important and should always form the basis for generating a forum for constructive dialogue among stakeholders along the value chains. This is because benching generates credible information for decision-making as sound prioritization and decision-making is difficult if parties are seen as advocating biased or unfounded positions or asking for special favours.

7.8 Improving Value Chain Financing

The diversity of geographic, environmental, agronomic, business and cultural factors across different countries of Africa and even within countries makes it impossible to recommend a single objective criteria in determining the appropriate financing arrangements for agricultural commodity supply and value chains. As such, a set of flexible subjective criteria by which a reasonable judgement about the relative challenges and risks presented by each cluster for finance could be assessed is provided here (Table 7.3). It is important to note that testing against these criteria does not substitute for an in-depth financial analysis of commodity value chains in the different countries. What is very apparent is that there seemed to be lack of well-executed financial analysis of the commodity value chains in Africa. Many value chain experts often have intuitive feelings about the relative profitability of commodity value chains, often underpinned by anecdotal evidence, but there is relatively little committed to paper in terms of how the finances of value chains in Africa operate in practice.

Table 7.3 Criteria for assessing value chain financing

Criteria	*Response*		*Comments*
	Yes	*No*	
Is there significant domestic consumer demand for the commodity in question?	✔		High local demand is a very positive factor
Is Africa a leading participant in export markets in relation to the commodity in question?		✔	Although there is domestic demand, the value chain should be internationally competitive to be able to compete with imports
Is the agricultural sector in Africa regularly exposed to major environmental (e.g. climate, disease, etc.) risks?	✔		There is the need to invest in technologies that curtail these risks (e.g. drip irrigation, disease resistant crop varieties, etc.)
Is the agricultural sector in Africa regularly exposed to unpredictable fluctuations in prices of either outputs or inputs?	✔		Investments in all year round production and strengthening of the value chain are required
Is the agricultural sector in Africa one in which there are significant opportunities for value addition through processing?	✔		There is great potential for value addition and actors such that producers and processors need to be linked more effectively.
Is the agricultural commodity supply and value chain development in Africa specifically identified by continental agricultural policies as a strategic priority?		✔	No concerted efforts targeted at commodity value chain development with clear financing arrangements
Does the national agricultural strategy go on to identify objectives and targets specifically related to the commodity in question?		✔	The objective should be to self-sufficiency in that commodity and the target should be to increase current output levels by for example 20% every year for the next 5 years
Is there objective, reliable and timely market and production data available on the specific commodity in question?		✔	There must be a database on the specific commodity as it is difficult to develop and monitor finance strategy without this
Is there some useful financial data and analysis that can be used to build up a picture of the typical financial structure of value chain participants and the flow of cash between them?		✔	This needs to be undertaken, and head- line results captured

(continued)

Table 7.3 (continued)

Criteria	*Response*		*Comments*
	Yes	*No*	
Is there a significant cluster of producers with farm sizes of sufficient size to be not only viable, but also potentially profitable?	✔		Geographical clusters of agricultural commodity producers of varying farm sizes that are profitable exist across Africa
Are there processing companies engaged in agricultural commodity processing in the country concerned or even Africa?	✔		This is vital for off-take but the existing processing companies in many production centres are currently not operating
Is there a well-structured network of producer organisations spearheaded by a professional management and technical team?		✔	This is vital in providing industry insight and promotion
At grass-roots level, are there clusters of strong (i.e. independent, homogenous, sustainable) producer groups, which might support group lending?	✔		Lending to small producers is usually more successful through group lending schemes
Are there examples of large commercial farming operations?		✔	Can be suitable customers, intermediaries for supplier finance and important for structuring the value chains
Does the value chain currently support formalized relationships such as con- tract farming or out-grower schemes?		✔	Formalized relationships in the value chains are an indicator that the value chains are fundamentally profitable, and out-grower /contract-farming schemes are good targets for finance
Is there a strong extension service available for producers?		✔	Robust extension service scheme is important, as financiers like to feel producers are supervised to a certain extent and receiving good advice
Do processors have access to high- quality Business Development Services?		✔	Support to the business and management skills of agribusiness is important to financiers
Is there potential for significant improvements in financial performance of specific market participants (e.g. producers, marketers or processors) through the adoption of new technologies or approaches?	✔		There is potential for significant productivity improvements given the appropriate investments

When examining a proposition, some of the key concepts of value chain analysis come easily to financiers, who are accustomed to looking at the strength and stability of a customer's supply chain, as well as their customer base. Businesses that operate within unpredictable and chaotic environments are unlikely to be viewed favourably. Thus the criteria for assessing commodity value chain financing adapted from GIZ (2011) are presented in Table 7.3.

7.9 Opportunities for Improvements

7.9.1 *Training of Farmers*

Farmers need training on the appropriate packages of good agronomic practices (GAPs) and application of inputs. This could be done through the provision of tailored extension services specific to targeted commodities. This requires that agriculture ministries be well resourced to be able to effectively execute their mandate.

7.9.2 *Training of Input Suppliers*

Majority of the input suppliers across different countries do not have adequate training and yet farmers rely on their advice and judgement on application of inputs. Building the capacity of input suppliers to be able to provide basic information on inputs and inputs applications is therefore very important.

7.9.3 *Certification of Seeds*

Most input suppliers currently import seeds from out of Africa and this makes seeds expensive to farmers. Thus opportunities exist for local seed production and this requires the training of producers. Training seed producers and certification of their seeds are important because seed production requires technical skills for optimum results. The guidelines and action points that meet international best practices in the seed value chain should be taken seriously.

7.9.4 *Strengthen Credit Linkages*

Financial institutions across countries must be willing to partner with farmers and other actors along the value chains to be able to present bankable business plans. However, this requires continuous follow-up with the financial institutions and negotiating a better deal/products for the collective/individual farmers and other commodity supply and value chain actors.

7.9.5 *Formation of Robust FBOs*

Some farmers are currently part of some form of FBOs. However, they do not use this platform for collective marketing of their produce. Opportunities exist for farmers to form robust FBOs that could be used to price their produce by leveraging on the collective platform for aggregating their produce and bargaining better deals with wholesales/brokers. Marketing through FBOs will help producers gain bargaining power to negotiate better deals for members. Thus FBOs should be located close to producer collectives and be made attractive as a trading intermediary for the farmers.

7.10 Cross-Cutting Issues

7.10.1 *HIV and AIDS*

HIV and AIDS can affect the performance of agricultural commodity supply and value chains but are currently not major issues of concern in the agricultural sector. People living with HIV and AIDS are still able to engage in various activities across supply and value chains with no major setbacks such as discrimination. This could be attributed to the success of HIV and AIDS sensitization activities implemented over time by national governments across Africa. Efforts by NGOs relating to advocacy and creation of enabling environment, coordination and management of HIV and AIDS as well as mitigation of social, cultural, legal and economic impacts, prevention and behavioural change communication with respect to National HIV and AIDS Strategic Frameworks have yielded the much-desired results.

7.10.2 Gender

The key gender issue affecting the overall performance of agricultural commodity supply and value chains is access to productive resources such as land and credit by women. In most of the clusters, women do not own land and mostly given marginal lands by their male counterparts to farm and because they are unable to access credit to invest in the purchase of inputs, they always record poor yields. Because of this, men generally dominate the production segment commodity supply and value chains with women controlling the marketing segments where they have the competitive advantage. Besides, the societal roles and responsibilities are heavily weighed against women and girls creating unbalanced power relations between men and women, which consequently influence access and control over resources such as land, credit and technology; access to time, markets, information and participation in decision-making at the household and community levels. These greatly affect the participation of women across the different segments of commodity supply and value chains.

7.10.3 Environment, Climate Change and Green Economy

Environmental degradation and climate change greatly affects productivity of commodity value chains. Soil fertility levels are low because of degraded lands and farmers have to rely on inorganic fertilizers and other agrochemicals to boost production and these create environmental sustainable challenges in the long term. Poor rainfall patterns and underdeveloped irrigation facilities in production areas also greatly affect the performance of value chains. High temperatures also affect the production and transport to consumption centres. These push commodity value chains actors to engage in unsustainable production, conversion and distribution practices that present serious challenges to the creation of a green economy in Africa.

7.10.4 Vulnerability

Key vulnerability issues include floods of farms, especially during the rainy season, outbreaks of pests and diseases and volatility in commodity prices across seasons. These negatively affect investments of actors along the chains, as the risk is too high to encourage long-term investments in agricultural production. This is particularly so with farmers engaged in rain-fed production.

7.10.5 Population

Population growth and changing structure also affects commodity supply and value chains. The youth are not attracted to farming due to the low image accorded farmers and the aging farm population is unable to supply to meet the complex standards required by the growing middle class. The rural–urban drift is bad for agriculture, which is largely practised in rural Africa. There is also the issue of conversion of farmlands to residential areas due to urbanization. Although these present challenges, they also present opportunities for the expansion of production to meet the increasing demands occasioned by population growth.

7.10.6 Culture

The African culture provides some opportunities and bottlenecks for the agricultural sector. The main opportunity provided by the culture is that the consumption of many staples is well known in the African culture and therefore the skills and knowledge required for commodity production are consistent with our culture. The challenge however, is that the culture does not facilitate innovation in many commodity supply and value chains. Farmers are reluctant to change their way of doing things and always sceptical in accepting improved production technologies that appear inconsistent with their culture, knowledge and value systems.

7.10.7 Security

Economic security and stable political systems are critical for enhanced performance of commodity value chains. Actors along the commodity value chains must be able to go about their activities without fear and should be sure that their investments are safe. This should be supported by stable democracy and well functioning judicial systems that settle disputes arising from business and related activities. However, a considerable number of countries in Africa are not having stable democracies and rule of law, which have negative implications for commodity value chain development.

7.10.8 Disaster

Frequent fire outbreaks and flooding of major market centres affect the productivity of commodity value chains. In most cases farmers and traders lose their crops and goods during such disasters. Traders and input dealers are always at the losing end whenever there are market fires and floods and this negatively affects investments in the end side of commodity supply and value chains. Roads linking production areas with market centres always get destroyed during floods, at huge costs to individuals and nation states.

7.10.9 Water Security

Water security affects agricultural production greatly and this is worsened by the climate change menace. In some cases farmers are unable to get water to produce all year-round and this affects availability of farm produce at critical times. To ensure water security, there must be investments in the construction of irrigation facilities across different countries. Besides, efforts should be intensified to provide water technologies including breeding of drought resistant crop varieties.

CHAPTER 8

Value Chain Profiling in Practice—A Case Study

Abstract Value chain profiling is an emerging area, especially in African agriculture. Understanding how to undertake a commodity value chain is thus critical in the efforts towards transformation of African agriculture for inclusive and sustainable development in the continent. As such, we present in this chapter a practical application of commodity value chain profiling. We started by establishing the gaps in supply of the selected commodity and how these gaps are bridged. We went further to present the trends in production, imports, exports and inputs prices. How trade policies affect the commodity value chains is also profiled. We conclude the chapter by practically demonstrating how to profile commodity value chain actors and their activities using the leaky bucket method.

Keywords Actors · Ghana · Profiling · Tomato · Value chain

A.-R. Alhassan and M. Abunga Akudugu,
Supply Chain Management in African Agriculture,
https://doi.org/10.1007/978-3-030-54209-2_8

8.1 Tomato Production, Imports, Exports and Price Trends in Ghana

8.1.1 The State of Tomato Production in Ghana

The productivity of tomatoes in Ghana is low. Currently, the average yield per hectare under rain-fed conditions is 7.2 metric tons versus the achievable yield of 15 metric tons. Thus the country is only able to get 48% of the attainable yield. There are about 518,443 farm households engaged in tomato production out of which 32,944 are in the coastal zone, 265,469 are in the forest zone and 220,030 in the savannah zone (GSS, 2014). These farmers according to the Ghana Living Standards Survey Round 6 (GLSS6) report produce about GHC374, 992, 080 worth of tomatoes annually with the coastal zone accounting for 7%, forest zone accounting for 37% and the savannah zone accounting for the remaining 56% of the total worth. This means that only about 36% of what is produced is sold with the remaining 64% either consumed or lost through post-harvest handling.

In terms of sales value, about GHC135, 800, 020 is realized annually from total sales of tomatoes with the coastal zone accounting for about 0.4%, forest zone accounting for 80% and the savannah zone accounting for the remaining 19.6% of total sales (ibid). The implication is that post-harvest losses are very high in the coastal and savannah ecological zones. This as outlined in FASDEP II and METASSIP I could be attributed to poor nature of roads to production centres and inadequate market information leading to weak market integration between local, district, regional markets; low standardization and product differentiation in domestic markets (weights and measures; grades and standards); uncongenial environment for trading in local markets; and limited marketing extension for producers, traders and exporters. Also, issues of inadequate product and cluster development; poor quality and irregular supplies of raw materials to agro-processing enterprises; inadequate institutional arrangement to support commercial scale agro-processing and low patronage of locally processed products are all to blame for high levels of post-harvest losses being recorded.

8.1.2 Production Trends Compared with Regional Neighbours

Tomato is one of the popular vegetables commonly cultivated by small-scale farmers in Ghana (Osei et al., 2010) and many other countries across the West African subregion as an income-generating activity. Data from the Food and Agriculture Organization (FAO) of the United Nations (UN) suggest that between 1961 and 2013, Ghana recorded one of the highest increases in area put under tomato cultivation relative to her neighbours—Benin, Burkina Faso, Cote d'Ivoire and Togo. Whereas Ghana recorded an astronomical increase of about 1437% in area under tomato cultivation during the period (i.e. 3,000 ha in 1961 to 46,100 ha in 2013), her neighbours namely Benin, Burkina Faso, Cote d'Ivoire and Togo recorded lower percentage increases (Fig. 8.1).

Specifically, Benin recorded about 2114% increase (i.e. 1,800 ha in 1961 to 39,849 ha in 2013) during the same period with Burkina Faso recording about 686% increase (i.e. 140 ha in 1961 to 1,100 ha in 2013). Also, Cote d'Ivoire recorded about 725% increase (i.e. 400 ha in 1961 to 3,300 ha in 2013) whiles Togo recorded the least increase of about 115% (i.e. 700 ha in 1961 to 1,503 ha in 2013). This means that there

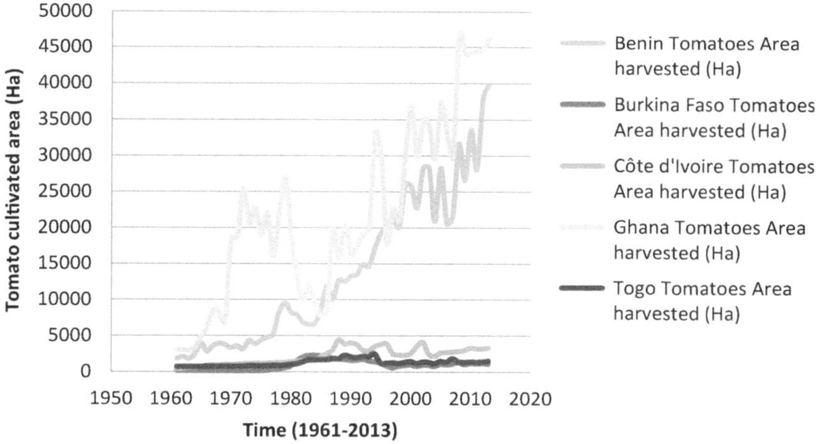

Fig. 8.1 Tomato cultivated area in Ghana and neighbours (*Source* Authors' computations based on FAOSTATS Data, 2015)

has been an increased cultivation of tomatoes partly due to the economic importance of the crop (Fig. 8.1).

The increases in area harvested of tomatoes in Ghana and neighbouring Benin, Burkina Faso, Cote d'Ivoire and Togo appeared to be in tandem with the trends in output over the period (Fig. 8.2). Data from FAO indicate that tomato output in Ghana increased from 15,000 tonnes in 1961 to about 340,218 tonnes in 2013 (2168%) compared to Benin, which recorded an output increase from 3390 tonnes to about 331,793 tonnes (9687%) over the same period. Within the same period, Burkina Faso recorded an increase from 1300 tonnes to about 11,500 tonnes (785%); Cote d'Ivoire from 4000 tonnes to about 34,734 tonnes (768%) and Togo from 2000 tonnes to about 6393 tonnes (220%). The data show that there has been a general increase in tomato production across West Africa with Benin recording the highest increase of output followed by Ghana with Togo recording the least increase in output levels. The high fluctuations characterizing output levels of Ghana and Benin should however, be of concern to policymakers, implementers and others.

The astronomical increase in area put under tomato cultivation and the corresponding increases in output across countries have only reflected marginally on yields. Although tomato yields have experienced a general fluctuating increasing trend between the period 1961 and 2013

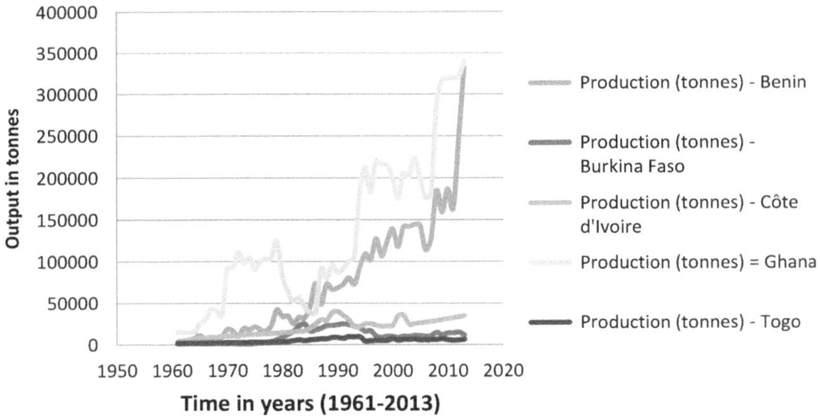

Fig. 8.2 Tomato production in Ghana and neighbours (*Source* Authors' computations based on FAOSTATS Data, 2015)

(Fig. 8.3), the yield increases do not seem to commensurate with the increasing acreages put under tomato cultivation over the period. In Ghana for instance, tomato yields according to the FAO (2015) data (Fig. 8.3) increased from 5.5t/ha in 1961 to about 8.1t/ha in 2013 representing about 47% increase. Comparing this yield increase with the increase in cultivated area (1437%) and output (2168%) indicates that increases in yields have largely lagged behind. It is reasonable to conclude based on the above data that the increase in output recorded is partly due to area expansion rather than productivity gains. This assertion is consistent with the observations by Akudugu et al. (2013) that the increases in output of staple crops in Ghana are due to area expansion than increased crop productivity.

Similarly, Benin recorded an increase in yields from 2.1t/ha in 1961 to about 9.2t/ha (338%); Burkina Faso from 10.2t/ha in 1961 to about 11.5t/ha in 2013 (13%); Cote d'Ivoire from 0.4t/ha in 1961 to 3.8t/ha in 2013 (850%) and Togo from about 3.1t/ha in 1961 to about 4.7t/ha in 2013 (52%). The fact that the increases in tomato output over the period are largely attributable to expansion in cultivated land area is worrying because this trend is not sustainable in the long term as land

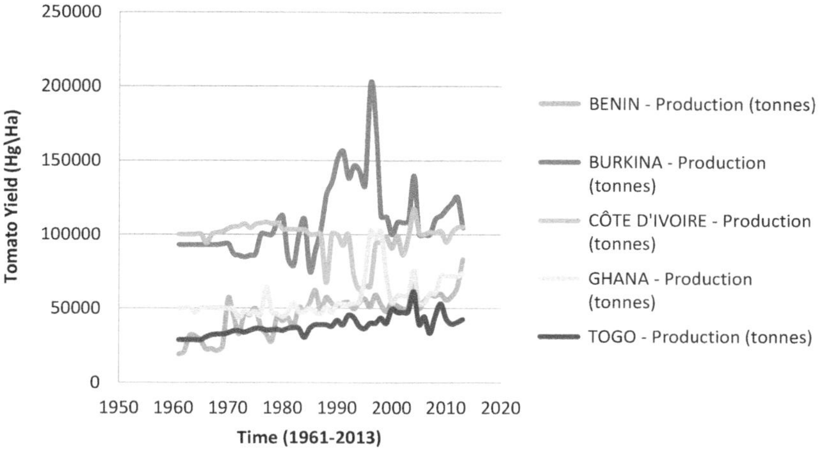

Fig. 8.3 Tomato yields in Ghana and neighbours (*Source* Authors' computations based on FAOSTATS Data, 2015)

resources are finite. The introduction of the supposed high yielding varieties (HYV) does not seem to have any significant effects on tomato output across Africa generally. There is the need for concerted efforts to increase productivity levels through the use of agri-smart technologies, particularly productivity enhancing farming methods.

8.1.3 *Imports and Exports Trends Compared with Regional Neighbours*

Despite the increase in tomato output recorded in Ghana since the early 1960s, the country continues to rely on imports to meet domestic demand. Available data (FAOSTATS, 2015) indicate that between 1961 and 2012, Ghana tomato imports (i.e. fresh tomatoes, peeled tomatoes and tomato paste) has increased by about 1798%, i.e. from 5000 tonnes in 1961 to 94,909 tonnes in 2012 (Fig. 8.4). The huge increases in tomato imports represent serious burden on the economy of Ghana contributing to the continuous depreciation of the Cedi.

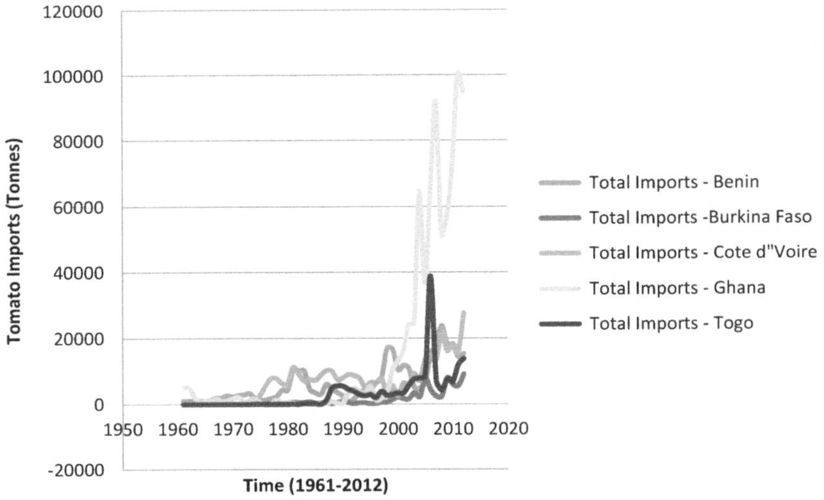

Fig. 8.4 Tomato imports in Ghana and neighbours (*Source* Authors' computations based on FAOSTATS Data, 2015)

The tomato industry in Ghana is complex and multifaceted. This means that halting and reversing the trend of reliance on imports to meet domestic demand obviously requires a multi-actor/sector approach, which can only work with the necessary political and institutional commitment. Just like Ghana, countries such as Benin, Burkina Faso, Cote d'Ivoire and Togo have also had astronomical increases in tomato imports over the period. For instance, Benin had its tomato imports between 1961 and 2012 increased by about 1436% (i.e. 1000 tonnes in 1961 to 15,355 tonnes in 2012); Burkina Faso moving from zero imports in 1961 to about 9052 tonnes in 2012; Cote d'Ivoire from 76 tonnes in 1961 to about 27,688 tonnes in 2012 representing an increase of 36,332% over the period and Togo also moved from zero tomato imports in 1961 to about 13,754 tonnes of imports in 2012 (Fig. 8.4). It must be noted that tomato paste and peeled tomato constitute the largest proportion of all imports across the countries examined. For example, about 97% of total tomato imports by Benin in 2012 was tomato paste; 99% of Burkina Faso's; 84% of Cote d'Ivoire's; 93% of Ghana's and 99% of Togo's tomato imports in 2012 was tomato paste.

In fact, even Nigeria, one of the largest tomato producing countries in the world moved from zero tomato imports in 1961 to about 147,043 tonnes of tomato imports in 2012 and almost all these imports (99.8%) were tomato paste (FAOSTATS, 2015). The data thus support the assertion that most countries in the West African Subregion and possibly Africa for that matter are not self-sufficient in tomato production, most especially value-added tomatoes such as tomato paste.

The good news is that at least some proportion of the produced tomatoes is exported. All the countries examined (Benin, Burkina Faso, Cote d'Ivoire, Ghana and Togo) moved from zero tomato exports in 1961 to some level of exports in 2012. Most of the countries started exporting tomatoes in the early 1990s, probably due to trade liberalization among other economic policies (Fig. 8.5). Specifically, Benin exported about 114 tonnes of tomato paste in 2012; Burkina Faso exported 14,183 tonnes of fresh tomatoes only in 2012; Cote d'Ivoire exported about 170 tonnes of tomatoes comprising 24% of fresh tomatoes and 76% of tomato paste; Ghana exported about 64 tonnes consisting of 89% of fresh tomatoes, 8% of tomato paste and 3% of peeled tomatoes and Togo exported about 81 tonnes of tomato paste in 2012 (Fig. 8.5). Interestingly, the trends in tomato exports across the countries examined seem to be moving together with common increases and declines.

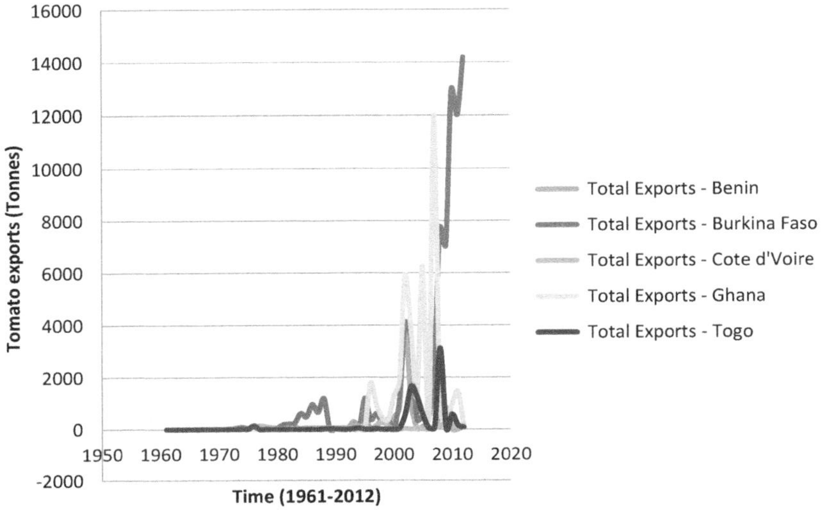

Fig. 8.5 Tomato exports in Ghana and neighbours

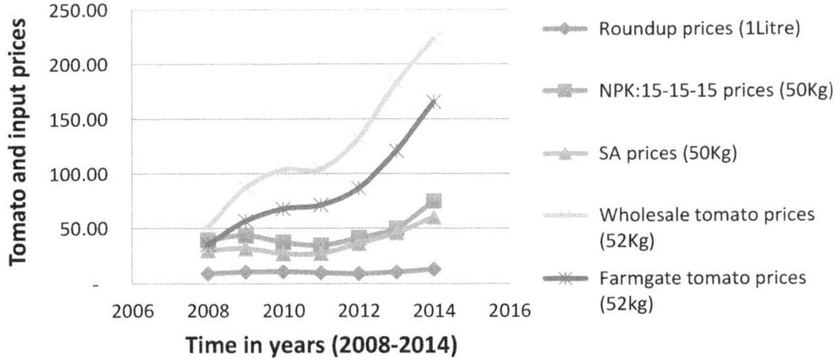

Fig. 8.6 Trends in tomato and agrochemicals prices

8.1.4 Trends in Tomato and Agrochemicals Prices

Since 2008, prices of agrochemicals for tomato production have been generally stable compared to output prices (Fig. 8.6). For instance, between 2008 and 2014, the prices of weedicides (roundup) and fertilizers (NPK and Sulphate of Ammonia—SA) only increased marginally. However, both farm gate and wholesale prices of a 52 kg box of tomatoes have increased exponentially over the same period (Fig. 8.6).

8.1.5 Ghana's International Trade Relations and Policy Objectives

International trade relations and policies influence the agricultural sector of Ghana and the tomato industry for that matter. These trade relations and policies as well as the policy objectives are discussed in the following subsections.

8.2 Actors and Activities Along the Value Chains

8.2.1 Input Dealers and Service Providers

- **Types of Inputs and Services**: The Agro-Inputs retailers such as Donewell Agro-Ventures, PAG Farms, Enesco Agrochemicals and Royal Ahomka AgroChemicals sell tomato seeds such as Technisem, Heinz,'Akoma', Pepto, Pagma, etc., to tomato farmers in the Wenchi District. Some of these seeds such as Heinz and Pagma were said to be exotic varieties, which are mostly imported from the Netherlands and France by agro-import companies. As observed by the agro-input dealers, some of the varieties after being purchased and used by farmers as a first filial generation seeds, those farmers extract seeds from them for replanting. The Agro-Inputs retailers also have fertilizers such as 15-15-15, 23-10-5, urea and ammonia sulphate to farmers. The Agro-Inputs retailers indicated that tomato farmers used mostly the NPK fertilizers (15-15-15 and 23-10-5) on their farms. This was confirmed when the researchers interviewed some of the producers and the District Director of Food and Agriculture in Wenchi. Also, the agro-input dealers sell Plant Protection Products (PPP) such as Lambda, which is a pesticide, and Margozeb, which is also a fungicide to farmers. Farmers also buy varied brands of weedicides/herbicides such as glyphosate, gallegan, sinco, etc., from the input dealers. Apart from chemicals, the input dealers also sell to farmers varied types of agro-equipment and protective gears such as knap sack sprayers, seed trays, water pumping machines, watering cans, wellington boots, nose masks, etc. Aside selling of agro-inputs to farmers, the input dealers, especially those who are registered like Donewell Agro-Ventures render extension services to farmers in the areas of chemicals handling, usage and dosage and safety precautions. The agro-input dealers were however quick to say that some farmers do not follow their advices in the application of these chemicals.

- **Sources**: The main sources of the these inputs, according to the agro-input dealers are the importers such as DW Dizengoff, Calli Ghana ltd, Chemico Ghana Limited, etc., and some major distributors such as Agyaaku Farms Training ltd, Woffa Addo AgroChemicals, Frankatson ltd, etc.
- **Availability, Accessibility and Affordability**: According to the dealers, the inputs are available from their supplies throughout the year and they are also accessible. They however mentioned that the affordability of the inputs is relative, on the side of the agro-inputs distributors the inputs are affordable but farmers do not see it so, as they complain of the inputs being expensive Mr. Samuel Kwame Fosu of Donewell Agro-Ventures said. The input dealers attributed the high inputs price to the hikes in petroleum prices and the dollar rate against the cedi, which affect transport and import fares (cost).
- **Sales Volumes**: The volume of sales depends on the season they sell averagely about 1000 bags of NPK, 600 litres of insecticides as well as 600 litres of fungicides in a week. A bag of NPK without a subsidy is GH¢120.00, the price of a litre of an insecticide is GH¢18.00 and that of fungicide is GH¢12.00. According to the input dealers they get higher sales in the rainy seasons with an average weekly sales of insecticides estimated to be about GH¢1500.00 which is about 9000 cartons, fungicides to be about GH¢ 300.00 which is about 1 carton and seeds to be about GH¢ 350.00 which is about 10 cartons.
- **Customers**: The main customers of the agro-input dealers in the Wenchi District are the individual smallholder farmers who are not only into tomato production but also into the production of other food crops. Aside the farmers, they also sell to other small-scale retailers or venders, outlet agents outside Wenchi District. Gardeners, farmer groups and associations like Mango Farmers Association and institutions like the Methodist University also buy agro-inputs from the agro-input dealers.
- **Enterprise Budget**: Most of the agro-input dealers visited were not able to tell the total cost of their goods and how much they spent on

transport and storage. Though they could hardly give their enterprise budgets, Mr. Samuel Kwame Fosu of Donewell Agro-Ventures gave the cost of a fertilizer consignment as GH¢1680.00, Transportation cost as GH¢160.00, but he could not tell rent, utility and storage cost, he however gave a rough estimation of the cost of the fertilizer consignment to be about GH¢2000.00. It is evident that most of the agro-input dealers do not keep coherent and up to date information about their enterprise budgets but if they do keep up to date records then it means that they are not willing to disclose their financial strengths to people they hardly know.

- **Value Addition**: On value addition, they said they do not add any chemical value to their inputs before selling to farmers. Some of them however said since the price of fertilizer is high, they sometimes sell in bowls and the only values added are quantity resizing and repackaging into smaller containers like plastic bags and fertilizer sacks. However, buyers often provide most of the containers.
- **Financial Support**: On financial support, the agro-input dealers indicated that they plough back their profits into the business and sometimes they get loans from microfinance and banking institutions. High interest rates are however the limiting factors to accessing loans from the banks and the microfinancial institutions the input dealers indicated. Also, they mentioned that some of the importers and the formulators of the chemicals do supply to them goods on credit bases because of the business relationship they have established with them over the years.
- **Other Services Providers**: Other services providers such as tractor services were not common in the district since the smallholder farmers' use hoe and cutlass in the cultivation of their tomatoes. But information from the District Director of Agriculture indicated that a cost of ploughing an acre of land is GHC80.00 just as manual labourers on contract also charge for making ridges.

1. **Group membership**: The research revealed that agro-input dealers belong to an association, which runs from the district through the region to the national level. Ideally, all agro-input dealers are supposed to belong to this group but some of them especially those in villages flout the directive. The name of the association is Ghana Agro-Input Dealers Association (GAIDA). It is an active

association and they meet once every quarter and also they meet on emergencies. There are 50 members of GAIDA in the Wenchi District, that is, 42 males and 8 females and the group is registered. The main activities of the association are to work together to protect the image of the agro-inputs industry by ensuring that members do not indulge in selling of fake or inferior inputs. Other activities of the association include sharing of knowledge and information regarding the inputs, helping members in times of need such as outdooring, weddings, funerals, etc. To become a member of the association an agro-input dealer must first register the business enterprise with Plant Protection and Regulatory Services (PPRS), Environment Protect Agency (EPA) and with certificates from the PPRS and EPA registration with GAIDA is then made. Members of GAIDA pay monthly dues of GH¢5.00, which helps them, finance the activities of the association at the district level. Some benefits derive from membership of GAIDA include disputes resolutions, shared integrity and credibility, knowledge upgrading through training from organizations like (IFDC), (GIZ) USAID (Advance) etc. Challenges that agro-input dealers face include farmers refusal to follow laid down procedures in handling and use of agrochemicals, presence of illegal agro-input dealers. As a group they lamented that the unregistered or non-members of the association who are into the sale of agro-inputs post a big challenge to the association and the entire agro-industry as they bring into the market substandard goods which they sometimes price lower to attract unsuspected customers.

- **Storage**: The research revealed that the input dealers have their own warehouses aside the shops where they store their stocks. Most of the inputs are stored on raised pallets above the ground with fertilizer being store in a curative bonding to aid air circulation between the bags. The period of storage is not fix or definite as according to them it is dependent on how long stock lasts which ranges from a week to even a year. Inputs that have short shelf life span are usually the most consumed commodities and are not stored for long.
- **Information Sharing and Payment of Taxes**: Information on pricing and other relevant issues is shared among the input dealers, suppliers as well as customers. In the case of the customers, they

share information through the extension services they provide to them at the point of sale. Some of the channels of communication are face-to-face discussions, phone calls, messages through people, e-mails, etc. Information is also shared during workshops, talk shows on radio and television, meetings, etc. The research also revealed that each agro-input dealer pays an average of GH¢50.00 as income taxes to Ghana Revenue Authority-domestic division and an average of GH¢15.00 as levies to the District Assembly annually.

- **SWOT Analyses**

  - **Strengths**: The input dealers saw their technical expertise as strength. Their credibility in dealing with good quality chemicals is also strength as well as having good and big warehouses so that they can store their inputs so as not to congest their stores with all their inputs.

- **Weakness**: Insufficient capital is their biggest weakness and also lack of collateral security to enable them source from the banks is a weakness.
- **Key opportunities**: According to the agro-input dealers, some of the farmers have formed associations and thus it is easier to meet with them when the need arises. Also, the major economic activity of the people in Wenchi and its surroundings is farming and this becomes a big opportunity for the input dealers to sell their produce. Membership to GAIDA was also identified as an opportunity through which their grievances are resolved. Government interventions like fertilizer subsidies are opportunities through which the agro-input business could grow as the distribution of the subsidized fertilizers is done through the input dealers.
- **Key threats:** The biggest threat to their business they said is rain and climate change, when there are no rains farmers will not buy inputs and that affects their businesses. The unregistered dealers also pose threats to the inputs business as customers turn to buy more of their cheap inferior products.
- **Potential Activities**:

  - **Sustain strengths**: The input dealers constantly upgrade their knowledge to provide good services to their clients. They always supply quality goods and give advices to their customers to maintain the confidence imposed in them.

- **Address weaknesses**: Profits from their sales is plough back into business so as to keep them in business. They always try to appeal to farmers' conscience to do the right thing by following inputs guidelines.
- **Minimize threats**: Though they said they could not do anything about rainfall pattern, they are into the sale of varied inputs which irrigation equipment so that they can still make some sales if rains are not good.
- **Activities to take advantage of opportunities**: They said that sometimes they use the radio stations to advertise their products. Dealers also always work hand-in-hand with government agencies like Ministry of Food and Agriculture (MOFA) to take advantage of government initiatives that involve the supply of inputs to farmers. The agro-input dealers also take advantage of trainings organized by MOFA/DADU and other organizations like IFDC, USAID, GIZ, etc., to upgrade their skills.

- **Inputs–output map for Wenchi Input Dealers**

  The top of the leaky bucket indicates the sources of income for the agro-input dealers hence their customers while the bottom indicates expenses (costs) of the agro-inputs. The percentages were calculated from total monthly cost (expenses) of fertilizer, taxes/levies and dues only, as input dealers did not give the monthly costs and sales of inputs and other transactions costs (Fig. 8.7). The profit as presented herein was calculated by subtracting total monthly costs (expenses) from total monthly sales.

8.2.2 Producers

- **Scale of production—Farm Size, Quantity Produced and Value added**: The average farm size of about 800 tomato smallholder farmers in the Wenchi District is about 1.5 acres. According to the farmers an average tomato yield per acre is about 10 boxes with all sizes (size; 52 kg, 120 kg or 190 kg) inclusive. The farmers said the yield is dependent on factors such as the type of variety, pests and diseases prevalence, agronomic practices, soil fertility and rainfall pattern of a particular production season. Producers said they add no physical value to the produce except for sorting them into

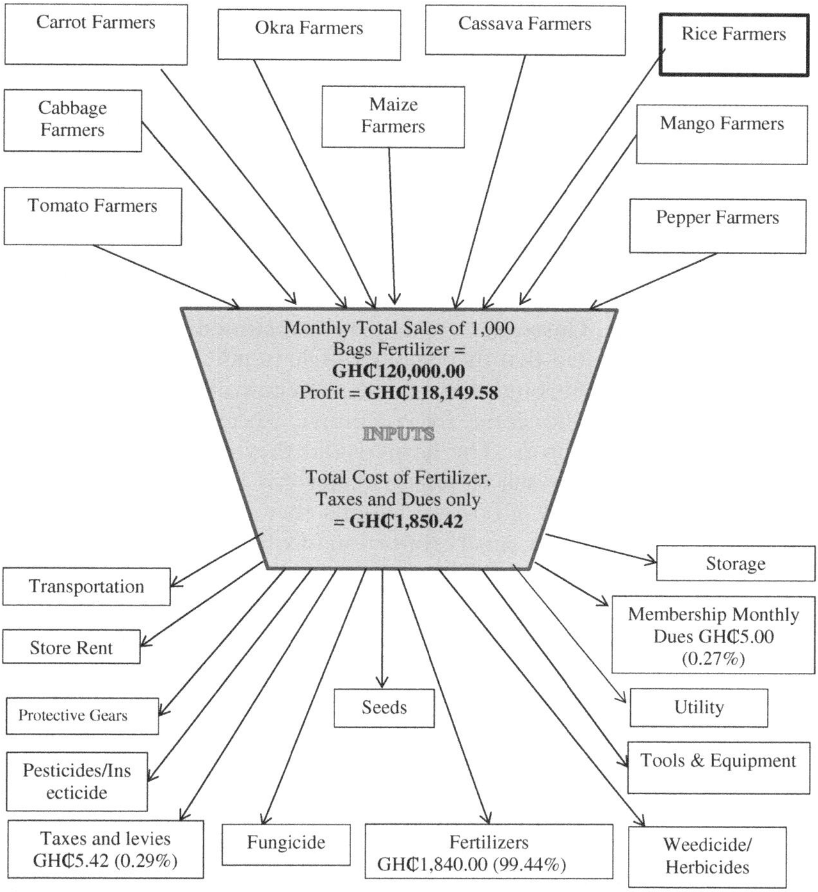

Fig. 8.7 Inputs–output map for Input Dealers in Wenchi

damage tomatoes and also grading them into fairly uniform sizes before selling to the marketers.

- **Quantity Sold and Amount Realized**: They sell almost all the produce except for a little amount that is consumed at home, which is often of the least quality. They could not readily tell how much harvest they had for the first season of 2015 due to rain failure but one of the producers by referring to previous harvest said the average harvest per acre was about 23 boxes, all sizes inclusive and all were sold except for one box. The farmer said each box of 52 kg was sold GH¢120.00, the 120 kg at GH¢170.00 and the 190 kg at GH¢250.00. He said the 52 kg boxes were 3, the 120 kg were 13 boxes and the 190 kg boxes were 6, and he had a total sale of about GH¢4070.00 from the 22 boxes. Prices of tomatoes are not fixed therefore producers and marketers negotiate on prices and farmers get higher prices for their produce during the lean production season they however receive lower prices for their produce during the peak of a production season.

- **Mode of Sales, Customers, Quantity Consumed and Given Out**: Producers indicated that their mode of sale is individual, they do not sale in groups or through FBOs. The main customers are the wholesalers, retailers who come from Kumasi, Accra and Wenchi who also act as aggregators. The farmers said they do not have specific customers that they sell to any marketer can buy from any farmer. Tomatoes are hardly given out as gifts they consume some of the tomatoes but in a very small proportion of what they sell. They could not however quantify the quantity consumed as it was not measured and consumption too was in bits but an approximation of about a box for a whole season was given as a quantity consumed and given out. Apart from the smallholder tomato farmers, a privately owned company tomato farm was also visited; the farm, a subsidiary of Tomacan a Wenchi based tomato processing factory was established to produce to feed the factory.
- **Farm Size and Capacity:** The total cultivated land size of the company farm is about 2.780 hectares (ha) with an open cultivated field of about 1.7 ha and the remaining land is used for 9 greenhouse facilities with each one measuring about 120 m^2 translating into about 1080 m^2 (1.8 ha). The capacity of the greenhouses is about 5 tonnes but currently they produce around 2.8 tonnes while the average yield of the open field is about 7 tonnes. The respondent was in the person of Mr. Osei John Jima added that averagely the farm produces about 10 tonnes in a season.
- **Value Addition, Quantity Sold and Amount Realized**: According to Mr. Osei John Jima who is the company farm production technician (Technical Advisor) tomatoes produce from the company's farm are not being processed, the only value they add to the produce are sorting, grading and packing into boxes for sale. When Mr. Jima was

questioned why the tomatoes are not processed he said the varieties produced are not industrial varieties because they have a lot of water and less pulp for paste and about 10 tons will yield 1 ton of paste. He further stated that the industrial varieties such as Heinz, Pepto and Pagma are being tried on the company's farm and they are also in talks with the Crop Research Institute (CRI) of the Council for Scientific and Industrial Research (CSIR) to develop more industrial varieties for the farm. Mr. Jima who stood in for the farm manager indicated that depending upon the amount of tomatoes produced by the generality of farmers in the area, between GH¢12,500.00 and GH¢21,000.00 is sometimes realized from the sale of the 10 tonnes of tomatoes. The farm mode of sale is individual, they do not sale in groups or through FBOs, as there are no FBOs.

- **Mode of Sales, Customers, Quantity Consumed and Given Out**: According to the production Technician of the company's farm about 80% of their produce goes to the wholesalers in Accra and Kumasi while the remaining 20% goes to the open-air market retailers. The main customers of the company's farm are usually wholesalers, retailers who are also aggregators, hotels, restaurants and institutional consumers like the Wench Senior High Schools who come from Wench, Techiman, Kumasi and Accra. When he was asked if the company is willing to sell through FBOs, Mr. Jima replied 'The aim of the farm is to produce to feed the processing factory (Tomacan), so if everything is set the farm may stop selling fresh tomatoes but processed them for sale'. He further stated that though there is no tomato FBO in Wench, the company is willing to sell through FBOs. On quantity consumed and/or given out, it was revealed that about 20 kg is shared among farm workers for consumption.
- **Production inputs and services**
  - **Types of Inputs and Services**: The major inputs used by tomato producers, both the smallholder farmers and the company farm are the tomato seeds which were once hybrids but due to replanting of extracted seeds and crossing with local varieties new hybrid varieties like 'Akuma', Technisem, have evolved but the exotic breeds such as Heinz, Pepto and Pagma are also cultivated. Fertilizers such as N.P.K (15-15-15 and 23-10-5) are mostly used and insecticides such as lambda,

confidon are also used. Weedicides like glyphosate, gallegan, sinco are used and fungicides such as Margozeb are also used. Farmers do receive extension services and training from the District Agriculture Development Unit (DADU) and that of input dealers.

- **Sources, Availability, Accessibility and Affordability**: The main sources of inputs used in the production of tomatoes in the Wenchi District are the agro-input dealers. According to the producers the inputs are available in the agro-inputs shops. The inputs are also accessible they are however very expensive especially fertilizer the farmers remarked.
- **Suppliers**: The agro-input dealers such as Donewell Agro-Ventures, PAG Farms, Enesco Agro-Inputs, etc., in the Wenchi District are the main suppliers of agro-inputs to tomato producers in the District. The company farm however indicated that they source most of their inputs from DW Dizengoff, Calli Ghana ltd and Chemico Ghana Limited in Accra and Tema. The representative of the farm manager Mr. Osei John Jima said they source the inputs from Accra and Tema because they feel the input dealers in Wenchi do compromise quality for cheapness. He admitted that prices of the inputs from their suppliers are slightly higher than those sold by inputs venders in Wenchi town. On availability of the inputs he said inputs are available but sometimes scarce as the inputs are often brought in on request. To the company farm, the inputs are accessible but costly he retorted.
- **Financial Support**: The smallholder producers said they self-finance their farming activities from their savings. However, Mr. Osei John Jima said the company's farm self-finances the production from produce sales but it also gets financial support from AGRA, COVET, EDAIF and expatriates.

• **Group Membership**: The tomato producers in the Wenchi are not into any association this was revealed when the research team interviewed the District Director of Agriculture Mr. Emmanuel Osei Adade. The director mentioned that even though there are a number of FBOs in the district, they were not aware of any FBO specifically formed for tomatoes production. This was confirmed with the farmers themselves as they said there are FBOs in the Wenchi

District but not specifically for tomatoes except for mango growers. They attributed the absence of FBOs to the difficulty in organizing themselves into one umbrella body.

- **Storage**: According to the smallholder producers their produce are not stored, marketers are called to come and buy whenever the produce are ready to harvest. Sometimes the marketers rather come around with their boxes looking for tomatoes to buy. The company farm respondent said that though they do have cold storage facilities, they store their produce in ordinary rooms with good ventilation. To store, yellow/partially ripe fruits are harvested and stored, this increases the shelf life up to a maximum of days or one week but the fully ripe fruits (reddish in colour) are often stored for two days.
- **Information sharing on inputs, markets/prices**: Farmers said they share information among themselves through informal channels of communications on inputs cost/prices, promotions, availability and sometimes inputs quality, production technologies and problems. Information is also shared among producers and marketers on produce availability, prices among others. Input dealers also share information with producers through the extension services they render as to how to use agro-equipment and chemicals.
- **Capacity to increase production to meet high demand**: Farmers said they have land and skills to produce more tomatoes but finances to expand their farms are lacking.
- **Willingness to engage in contract farming and requirements**: Farmers were divided on the issue of contract farming, while some of them will want to be engaged in contract farming if terms and conditions will be good for them others utterly rejected it what so ever the conditions will be. The proponent group of farmers for contract farming explained that sometimes they do get into agreement with their customers who do provide them with some assistance which they later pay in kind with their produce. The opposing group of the farmers however has the view that contract farming makes them vulnerable to be cheated by companies who engage them in the contract farming.
- **Modes of transportation of inputs and outputs and its reliability**: The only means for the transportation of inputs and tomatoes in the district is by public transport such taxis, 'Trotro' cargo trucks and sometimes by personal motorbikes. These transportation systems are very reliable but for the produce the marketers normally arrange with

transporters to come to the farm gates for their tomatoes therefore the farmers are not directly involved in the transport of their produce.

- **Payment of taxes and levies**: The smallholder farmers do not pay levies/taxes directly but they believe the input dealers pass on taxes to them in the form of inputs prices. The company farm is however registered and therefore pays about GH¢120.00 as taxes and levies to the Ghana Government through the Ghana Revenue Authority and the District Assembly.
- **Buyers of produce and quantities by buyers' types (%)**: According to Mr. Jima the company farm sells about 80% of its produce to supermarkets like Marina Mall in Accra and the remaining 20% is sold to open-air market wholesalers, retailers and institutional consumers. The smallholder farmers on the other hand indicated that 80–90% of their produce goes to the middle-women (wholesalers) who aggregate the produce from the various farms to the market centres. The remaining 10–20% is shared among roadside retailers and open-air market retailers. They further stated that individual consumers often buy from the roadside and open-air market retailers therefore they do not deal directly with individual consumers.
- **Input Requirements: Both labour and Non-Labour (per acre)**
  The inputs requirements of both labour and non-labour to produce one acre of fresh tomatoes in the Wenchi District are shown in the following tables. The tables are presented as Table 8.1 a on smallholder producers labour requirements per acre and Table 8.2 for company's farm labour requirements per acre. Table 8.3 is presented on smallholder producers' non-labour requirements per acre while Table 8.4 is presented on company's non-labour requirements per acre.
- **Intra and interrelationships**: There has not been any formal relationship among the tomato producers however there is a business relationship between them, between them and the buyers as well as between them and the agro-input dealers. This relationship is based on the trust that exit been them which they have built over the years.
- **SWOT Analysis**:

  - **Key Strengths**: The farmers said they have available land and technical expertise to produce more tomatoes. They also have

Table 8.1 Labour requirement (per acre) for smallholder farmers in Wenchi

Main activity		*Family Labour*	*Hired Labour*		*Total Cost*	*Tractor/other*	
			Quantity/№	*Unit Cost (GH¢)*		*Qty*	*Unit Cost (GH¢)*
Land preparation	Ridging (by day)	–	4	15.00	60.00	–	–
Raising seedlings		–	4	60.00	240.00	–	–
Watering	Hand watering	–	2 (2x/wk)*4	15.00	240.00	–	–
	Rainfed	–	–	–	–		
Weeds control	Spraying	–	4	20	80	–	–
	Earthling up	–	4	100.00.	400.00		
Pests & diseases control	Spraying	–	4	20	80	–	–
Staking		–	–	–	–	–	–
Mulching		–	–	–	–	–	–
Harvesting		–	4	60.00	240.00	–	–
Total		–	–	290.00	1340.00	–	–

the credibility to attract more buyers if more tomatoes are produced the farmers remarked.

- **Key Weaknesses**: Farmers identified their weaknesses as being their inability to expand the farms buy modern equipment due to insufficient capital.
- **Key Opportunities**: The district abounds with good road networks that connect the district to other urban centres like Techiman. The farmers also indicated that the peaceful nature of the district is an opportunity for them to concentrate on their farming businesses and to attract investors. The availability of input dealers was also identified as an opportunity for their production. The presence of the Tomacan factory couple with the large market availability is also seen as opportunities for their production. Government initiatives like subsidized fertilizers are opportunities they noted.
- **Key threats**: Farmers identified insufficient/erratic rainfall and climate change as major threats to their tomato businesses. The producers noted that high cost of inputs coupled with pests and

Table 8.2 Labour requirements for the company farm in Wenchi

Main activity		*Family Labour*	*Hired Labour*		*Total Cost*	*Tractor/other*	
			Quantity/№	*Unit Cost (GH¢)*		*Qty*	*Unit Cost (GH¢)*
Land preparation	Ploughing	–	–	–	–	1	80.00
Raising seedlings		–	4	60.00	240.00	–	–
Watering	Hand watering	–	2 (2x/wk)*4	15.00	240.00	–	–
	Rainfed	–	–	–	–		
Weeds control	Spraying	–	4	20.00	80.00	–	–
	Earthling up	–	4	100.00.	400.00		
Pests & diseases control	Spraying	–	4	20	80	–	–
Staking		–	–	–	–	–	–
Mulching		–	–	–	–	–	–
Harvesting		–	4	60.00	240.00	–	–
Net Total		–	–	275.00	1280.00	–	80.00
Grand Total				355.00	1280 + 80 = 1360.00		

Table 8.3 Non-labour inputs requirements (per acre) for smallholder farmers

Main Activity	*Cost item*		*Quantity*	*Unit cost(GHC)*	*Total cost(GHC)*
Seeds/ seedlings	Seeds		100 gms	30.00 (50 g)	60.00
	–		–	–	–
Fertilizer	15-15-15		2 bags	120.00	240.00
	Ammonia sulphate		1 bags	110.00	110.00
Watering	Rainfed		–	–	–
Chemicals for control of weeds, pests and diseases	Insecticides	lambda	2 litres	35.00	70.00
	Fungicides	Margozeb	3 litres	35.00	105.00
	Weedicides,	Gallegan	2 litres	35.00	70.00
		Sinco	2 litres	70.00	140.00
Sticks & ropes for staking	–		–	–	–
Material for mulching	–		–	–	–
harvesting Tools	Hand picking and buckets		–	Respondent could not recall	Respondent could not recall
Packaging and marketing	Plastic trays and wooden boxes		40	Respondent could not recall	Respondent could not recall
Total	–		–	405.00	735.00

diseases prevalence seriously threaten their farming businesses, they also indicated that indiscriminate use of agrochemicals coupled with declining soil fertility is posing serious threats to their tomato businesses. Farmers indicated that absence of tomato FBOs in the district affects their farming activities in the areas of pricing their produce and accessing bank loans as the banks are unwilling to give loans to individual farmers.

- **Potential activities to**:
  - **Sustain strengths**: As stated by farmers, they continue to learn and adopt new technologies to be abreast with modern production skills. Farmers said 'We are always loyal to our customers in order to maintain our credibility'. The farmers also said they acquire more lands for expansion and production increase.
  - **Address weaknesses**: In order to address their weaknesses, farmers said they are selective in the inputs they buy. Also, they are in the process of forming active association/FBOs to be

Table 8.4 Non-labour inputs requirements (per acre) for the company farm

Main Activity	*Cost item*		*Quantity*	*Unit cost(GHC)*	*Total cost(GHC)*
Seeds/ seedlings	Seeds		100 gms	(30.00/50gm*2).	60.00
	–		–	–	–
Fertilizer	15-15-15		2 bags	120.00	240.00
	Ammonia Sulphate		1 bags	110.00	110.00
Watering	Rainfed		–	–	–
	Irrigation	pumping machine	1	7000.00/5yrs = 1400.00	7000.00
		Drip kits	1	5749.00/5yrs = 1149.80	5749.00
Chemicals for control of weeds, pests and diseases	Insecticides	lambda	2 litres	35.00	70.00
	Fungicides	Margozeb	3 litres	35.00	105.00
	Weedicides	Gallegan	2 litres	35.00,	70.00
		Sinco	2 litres	70.00	140.00
Sticks and ropes for staking	Bamboo Sticks		1900	Could not recall	Could not recall
	Binding wire		10 rolls	50.00/5yrs = 10	100.00
Mulching	–		–	–	–
harvesting Tools	Hand picking and buckets		–	respondent could not recall	respondent could not recall
Packaging and marketing	Plastic trays and wooden boxes		40	respondent could not recall	respondent could not recall
Total	–		–	2994.80	3444.80

NB 7000/5 or 50/5 and means unit cost of the respective item (7000.00 or 50.00) divided by the average life span of the item (5 years) to get yearly/production season cost of the item (1400.00*1 = 1400.00 or 10 *10 = 100.00); where 1 or 10 is the quantity of the item needed

able to determine prices of their produce and to sell through the FBOs.

- **Minimize threats**: According to the smallholder farmers they are collaborating with Council for Scientific and Industrial Research (CSIR) to acquire climate smart varieties. The company's farm is also adopting the use of greenhouse and irrigation systems to minimize climate change effects.
- **Take Advantage of opportunities**: Producers said they take advantage of opportunities like government fertilizer subsidies, irrigation schemes, new varieties release by CSIR through the

District Agriculture Development Unit (DADU) and the extension services provided by the district Ministry of Food and Agriculture (MoFA) staff. They said they also take advantage of the available large tomato market to produce more tomatoes.

- **Inputs–output map for Wenchi District Tomatoes Producers (per acre)**
  There are two leaky buckets for tomatoes producers in the Wenchi District, one is for smallholder farmers and the other one is for a company farm (Fig. 8.8 and 8.9). The top of the leaky buckets indicates the sources of income for the producers hence their customers while the bottom indicates expenses (costs) of the agro-inputs. The percentages were calculated from the monthly inputs cost and monthly sales (Figs. 8.8 and 8.9). The producers do not sell to only customer type and also along the distribution chain they are many customers who buy from multiple sources except for the wholesalers, open-air market retailers and supermarkets, hence the reason for not attaching figures and percentages to the customers. The profit as presented herein was calculated by subtracting total monthly costs (expenses) from total monthly sales.

8.2.3 Marketers

- **Marketer type**: The marketers were mostly wholesalers, open-air market/roadside retailers and grocery stores.
- **Market Access**:
  - **Sources, Availability, Accessibility and Affordability**: Marketers buy tomatoes from tomato farmers in the Wenchi town and its surrounding communities such as Nkonsia, Awusa, Nkyera, etc. In terms of availability, tomatoes marketers get tomatoes all year-round except during the lean production periods, which usually fall between December, January and February when the rains are less. During these periods they go as far as to Burkina Faso for tomatoes. Tomatoes are also accessible in the district except for the lean production periods when they have to travel to neighbouring Burkina Faso to buy tomatoes. In terms of affordability, they said the tomatoes are expensive during the lean production periods as box, the

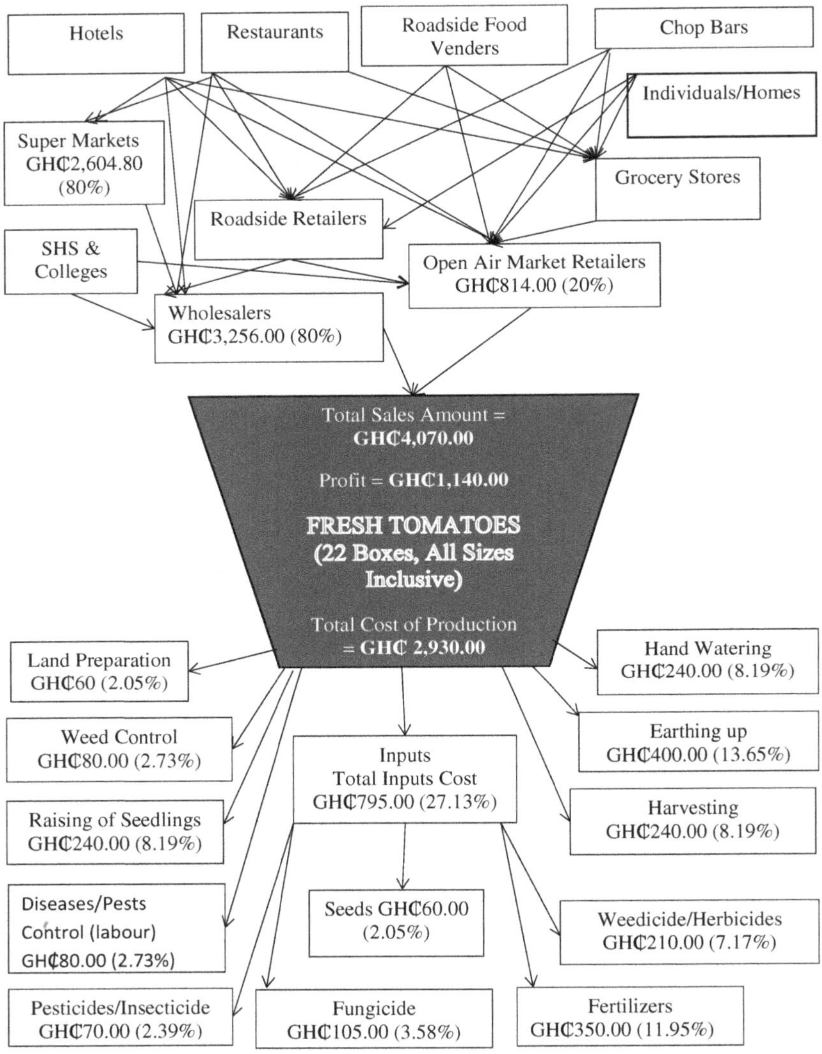

Fig. 8.8 Inputs–output map for smallholder tomato producers in Wenchi

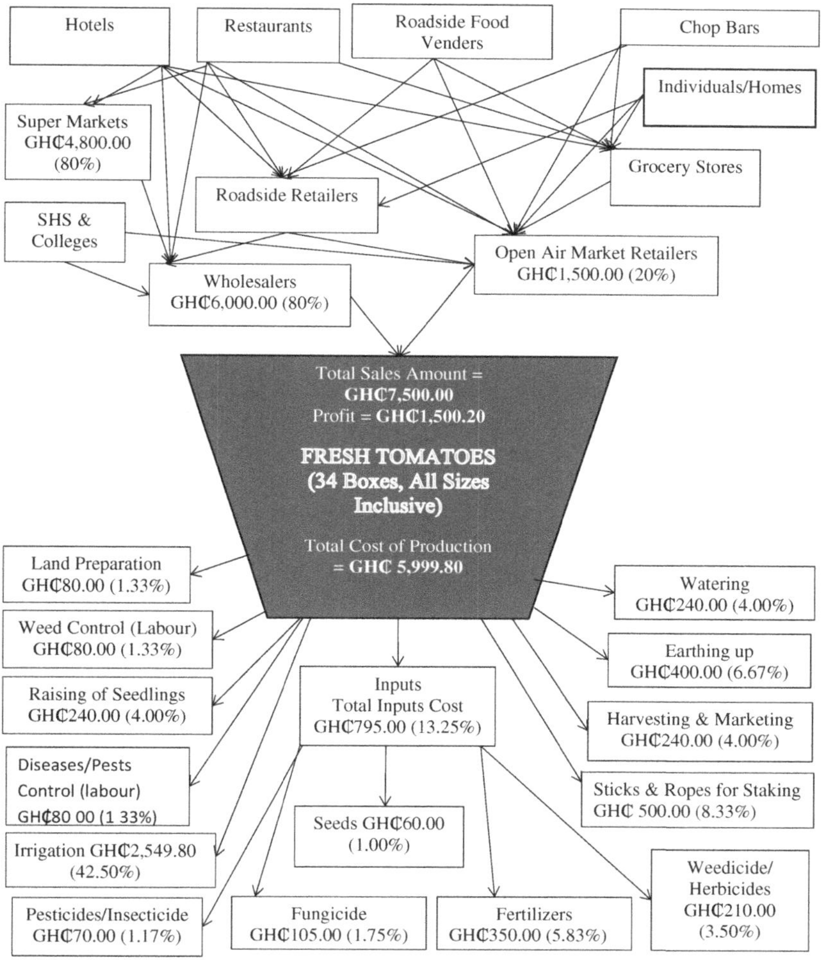

Fig. 8.9 Inputs–output map for Tamacan Company Tomato Farm in Wenchi

medium size (120 kg) can cost GH¢350.00 however tomato prices can come down to GH¢150.00 for the 120 kg box during the peak of harvest or production. Also, the marketers explained that prices of tomatoes also go up whenever Accra and Kumasi marketers come to buy because they offer higher prices for the commodity.

- **Sales Volumes and Customers**: Through the research it was revealed that an average open-air marketer can sell up to 2 boxes of any of the types of the boxes (size; 52 kg, 120 kg or 190 kg) in a week and can make up to GH¢50.00 profit per box. Main customers of the marketers are individuals, institutions such as Wench Senior High School, chop bar operators, food vendors open-air retailers and grocery stores.
- **Enterprise Budget**: On the enterprise budget, they said an empty box is charged GH¢2.00 while a box with tomatoes is charged GH¢10.00 as lorry fares. Loading boys take GH¢5.00 per box for loading the tomatoes into a truck and GH¢3.00 for off-loading them from a truck. They said a marketer can buy up to 80 boxes comprising of about 23 boxes of 52 kg, 42 boxes of 120 kg and about 15 boxes of 190 kg in a month depending on the availability of tomatoes and the financial capability of a marketer. According to the marketers they bought a 52 kg box of tomatoes at GH¢120.00, the 120 kg box at GH¢170.00 and the 190 kg box at GH¢250.00 and can go up during the lean production season. The average cost of the 80 boxes was estimated at GH¢20,400.00. To cover cost and to make some profit the marketers said they sold the 52 kg at GH¢184.00, the 120 kg at GH¢234.00 and the 190 kg at GH¢314.00 and the average amount realized for selling the 80 boxes in a month was estimated at GH¢28,080.00
- **Value Addition**: The marketers said they add no value to the tomatoes before sale except for sorting, grading and packaging in polythene bags (take away) for customers. They do not even wash the tomatoes and they are of the view that washing reduces the shelf life of tomatoes.
- **Financial Support**: Marketers said they support their tomatoes buying business through personal savings, they also go for loans from microfinance institutions but due to high interest rates they do not patronize their services unless in serious need. Some of them who have established a good relationship with farmers sometimes make part payment for their produce and later pay after sale.

- **Group Membership**: Marketers in the Wenchi belong to an association known as Wenchi Tomato Sellers Association with Mrs Akosua Fati as their president and Mrs Mercy Afrieye as their Secretary. The group is active as they hold regular meetings on a quarterly basis. According to the marketers the group has about 46 members with all of them being women the group is however not registered though the District Assembly recognizes it and usually invites its members for meetings at the assembly. The main activities of the group are prices negotiations, helping each other in times of need such as funerals, outdooring and other social issues. The basic requirement to join the association is to be a tomatoes marketer and also ready to pay an entrance fee of GHC100.00. The group meets every three months and members pay monthly dues of GHC5.00 and that is what is used to support the activities of the group. Sometimes money is raised on silver collection from members as and when there is the need to do so. The benefits are enormous which include social support; members also come together to decide on the prices and its accompanying quantities of tomatoes as and when the prices of tomatoes at the farm gates go up. According to members of the Wenchi Tomato Sellers Association the group faces many challenges, which include members not willing to pay dues, tomato price fluctuations and inadequate funds to support its work, which includes registration at the Registrar General Office in the Department Cooperative Societies.
- **Storage**: There are storerooms built by the assembly for market women, four marketers occupy each store. Products are stored with other vegetables at a fee of GH¢ 2.00 for a week. These stores are not cold stores a marketer retorted. Tomatoes are stored in baskets for good aeration to reduce the rate of deterioration.
- **Information sharing**: The group meets to share information among members when there are prices hikes, and also to decide on the acceptable prices to negotiate with producers. Information is shared through phone calls and face-to-face verbal communication the group president said.
- **Intra and interrelationships**: There is a cordial relation among colleague marketers they also have business relationship with producers and transporters. All these relationships are healthy among all the actors that the marketers work directly with.

- **Willingness to pre-finance producers**: They said it is a good idea but they do not have the money to pre-finance farmers.
- **Payment of levies and taxes**: The marketers pay taxes of GH¢0.50 daily even on weekends and holidays to the District Assembly revenue collectors. So long as they come to market to sell, they will have to pay to the assembly one of the marketers retorted.
- **SWOT Analysis**:

  - **Key strengths**: One of the strengths of the marketers is their ability to negotiate well with farmers. They are capable of travelling longer distances in search of tomatoes. They own the boxes hence sizes of the boxes are largely dependent on them.
  - **Key Weaknesses**: Marketers said they do not have their own means of transport and also are unable to fix prices for farmers. They also said they have no control over lorry fares as they do not take part in deciding on the fares they only negotiate with drivers or truck owners on fares to convey their goods.
  - **Key Opportunities**: According to the marketers the existence of large number of consumers coupled with the relative peace in the district is an opportunity for them to go about their farming businesses. The good roads coupled with large market are also seen as opportunities.
  - **Key threats**: The marketers in Wechi say their major threat is climate change which is threatening their business as the farmers are not getting good yields which make the available few boxes very expensive. They also said price instability is a challenge to them and also they face threats from other buyers from Accra and Kumasi as they tend to offer farmers' higher prices than they can do.

- **Potential Activities to**:

  - **Sustain strengths**: Marketers are constantly in touch with the farmers and always in for good price negotiations. The marketers explained that they try to always be patient with farmers so that they will not loss trust with them.
  - **Address weaknesses**: According to the president of the marketers association they are in constant talks with the Ghana Private Roads Transport Union (GPRTU) to include them in

deciding lorry fares. She said being able to negotiate well with farmers for good prices is the best way to go.

- **Minimize threats**: They said there is nothing they can do about the climate change, but for the external buyers they are to form an alliance with them to always agree on a common price any time they come to Wenchi to buy tomatoes.
- **Take advantage of opportunities**: One of the marketers said the market is an opportunity though they pay levies for using the market shades it is an opportunity for them to be easily located by their customers.

- **Inputs–output map for Wechi District tomato marketers**
  The top of the leaky bucket indicates the sources of income for the marketers hence their customers while the bottom indicates expenses (costs) of goods. The percentages were calculated from the overall total cost (Fig. 8.10). The marketers do not sell to only one customer type and also along the distribution chain they are many customers who buy from multiple sources hence the reason for not attaching figures and percentages to the customers. The profit as presented herein was calculated by subtracting total monthly costs (expenses) from total monthly sales.

8.2.4 Transporters

- **Transport types**: The major transport types in the Wenchi District are the private transporters who render services to the public. Traders do not have their own means of transport rather they use the private–public transport.
- **Availability, Accessibility, Affordability and Value Addition**: The transports are available, accessible but a bit expensive as the marketers always complain, the Chairman of the Wench Branch of the Ghana Private Road Transport Union (G.P.R.T.U) Mr. Kofi Sebend explained. According to him the cost of transportation is largely dependent on the cost of fuel and spare parts and because fuel and spare parts are expensive that explains why the transportation is a bit expensive. To access transport services to transport tomatoes one needs to book for a truck (s) and services are rendered on first come, first serve basis he said. According to the chairman

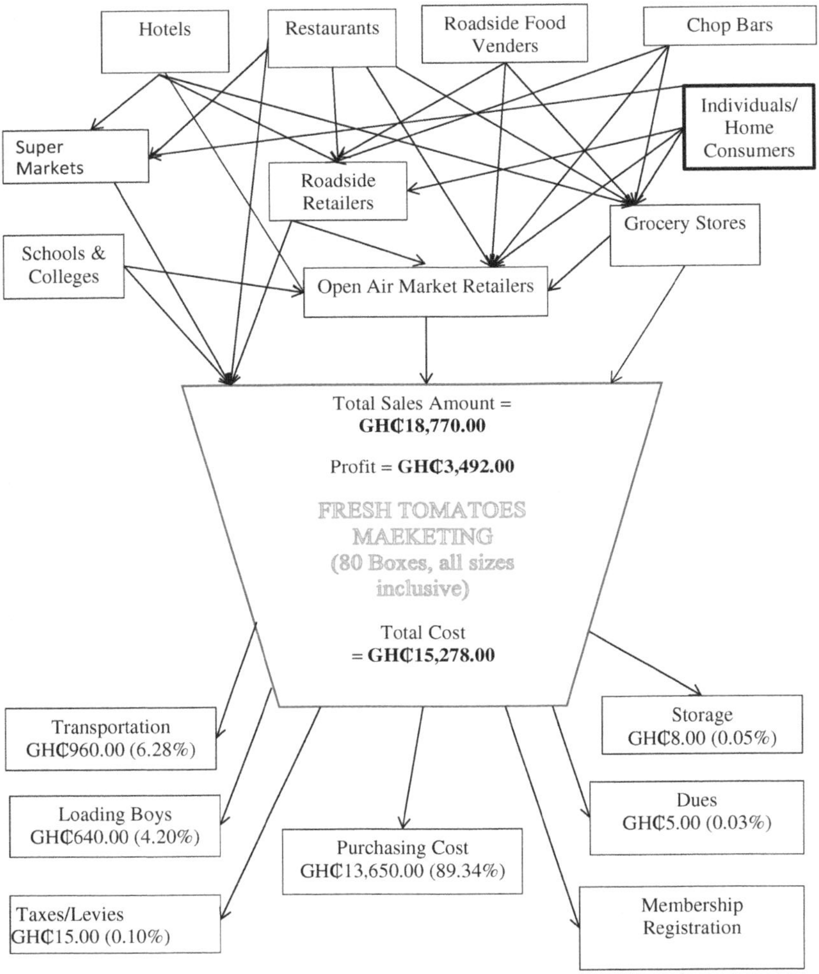

Fig. 8.10 Inputs–output map for Wenchi District tomato marketers

buyers from Kumasi and Accra come by their own means of transport but when that is not enough they seek for transportation from the Wench transporters to transport their goods. On value addition, the chairman said apart from loading, transporting and off-loading they do not add any value to the tomatoes.

- **Intra and Interrelationships**: When the chairman was questioned about their relationship within the transporters and with other customers of the transport services, he said they have good relationship among themselves as drivers and also they are in good business relationship with the marketers.
- **Information sharing**: On information sharing it was revealed that communication is formal and must be passed through the leadership of the organization. Information is shared among members regarding fares, prices of spare parts and other services. Between the transporters and marketers, information is mostly shared on fares and availability of services.
- **Willingness to pre-finance producers**: According to the chairman they do not pre-finance producers but they sometimes render services to customers especially marketers on credit basis for them to pay later. He however said it is not every farmer or marketer they render services to on credit basis except for those who are credit worthy.
- **Payment of taxes and levies**: The drivers of the union pay between GH¢10.00 and GH¢ 15.00 per load as loading levies to the District Assembly through the union (GPRTU) whenever they transport goods and the frequency of payment of the levy by a driver depends on how often the driver loads or transport goods in and out of the district. Drivers who transport goods on longer journeys, for instance from Wenchi to Kumasi, Takoradi or Accra often load twice or thrice in a week that is if the marketers do not come by their own means of transport the Secretary of the union Alhaji Baba Musah explained.
- **SWOT Analyses**

  - **Key strengths**: The union is a recognized registered organization and belongs to the mother union at the regional and national levels. Members of the union are law abiding according to the chairman of the union and that is strength. The chairman also identified the mandate of the union authority to sanction drivers for wrong doing as strength.
  - **Key weaknesses**: The union is unable to bring all drivers under its control, as some drivers operate without belonging to them

or any group. According to the chairman the union is weak in enforcing laws and order as it cannot prosecute recalcitrant drivers and cargo owners on its own.

- **Key Opportunities**: The chairman remarked that the ongoing road network upgrading in some parts of the district is a potential. Also there have been a number of trainings organized by Ghana Road Safety Commission (GRSC) and that were opportunities for the union members to upgrade their knowledge and skills he explained.
- **Key threats**: The secretary of the transporters union (G.P.R.T.U) Alhaji Baba Musah said higher prices of petroleum products, especially petrol and diesel are major threats to the union and the transportation industry in general. He also said, the fast depreciation of the cedi against the U.S. dollar is threatening their business because it negatively affects the prices of spare parts because importers pass on the cost to them. Bad roads were also identified as one of the biggest threats to the transport sector in the district as the bad roads cause rapid wear and tear on their vehicles therefore increasing the cost of maintenance.

- **Potential activities to**:
  - **Sustain strengths**: According to the secretary the union holds regular meetings to deliberate on issues such as disputes resolutions within the union and between the union and other players in the transport industry.
  - **Address weaknesses**: Alhaji Baba Musah said the union is in constant talk with drivers and car owners who are not with the union to bring them on board.
  - **Minimize threats**: The Union Secretary said transport fares are negotiated and increased to commensurate the increases in petroleum prices and spare parts.
  - **Take advantage of opportunities**: The union's members seize the opportunities like the trainings that are being organized by the Ghana Road Safety Commission are highly patronized by the union members.

8.2.5 Processors

- **Types of Processors and Customers**: Three categories of processors were identified in the Wenchi Township these categories were Home Processors, Small-Scale Commercial Tomato Mills and Cannery Factory.
- **Home Processors**: The home processing involves tomatoes being processed into various forms such as puree for stew or soup and chopped/sliced tomatoes for salad during food preparation in the various homes for family dinners.
- **Small-Scale Commercial Processors**: About seven (7) Small-Scale Commercial Tomato Processors were identified in the Wenchi Township and their activities include milling of fresh tomatoes into puree. These small-scale commercial processors do not buy tomatoes on their own to produce for the market they only render milling services to customers. Their customers include individuals, food venders, chop bars and restaurants.
- **Cannery Factory**: The only Tomato cannery in the Wenchi District known, as Tomacan is a privately owned factory, which stopped operating since 2007 this, was revealed by Mr Osei John Jima who was a production advisor to the Company Farm. According to Mr Osei John Jima the company stopped processing tomatoes because it was operating at a loss as about 10 tons of fresh tomatoes of the local varieties such as 'Akoma' and 'Technisem' could only be processed into 1 ton of tomato paste, this was due to the fact that the tomato varieties that were produced by farmers contained a lot of water with little flesh/pulp for paste. To address the problem the company brought in parano variety, which was an exotic variety that had less water but higher pulp/flesh. The variety was given to farmers to produce for the company but at the end of the production season farmers rather sold the produce to marketers from Accra and Kumasi as they offered higher prices to the farmers than what Tomacan could offer. Farmers gradually abandoned the production of the parano variety for the production of the local varieties ('Akoma' and 'Technisem') because the parano variety they claimed tasted a bit sour which marketers dislike as against 'Akoma' and 'Technisem' varieties which have alkaline taste. Mr. Jima further stated that another reason for farmers' refusal to adopt exotic varieties like Heinz, Pepto and Pagma, which are good for processing into tomato paste, is that the

varieties are susceptibility to diseases and have higher cost associated with their production.

- **Production Capacity and Source of Raw Materials**: The home processors said they buy tomatoes from the retailers, tomato-grinding mills stated that they do not buy tomatoes for processing it is their customers who bring their own tomatoes for grinding. According to Mr. Jima the company when was in operation could process up to 1000 tonnes in a week, tomatoes were brought from all over Brong-Ahafo and parts of Ashanti Region like Akomadan.
- **Value addition**: Apart from the Tomacan factory which was adding chemical value such as additives and preservative and physical value such as grinding into paste and packaging, the small-scale grinding mills and home processors only add physical value by grinding the tomatoes though sometimes other vegetables and spices like onions and peppers are added to the tomatoes to be processed into puree.
- **Group membership**: None of the processors identified were into any form of a group or association, not even the Tomacan factory belonged to any association. The processors said they are willing to be in a group, which they believe will provide a platform for them to air their voices, and to also harmonize their activities.
- **Main Challenges**: Seasonality of raw materials is a big challenge to them, the processors indicated. According to the respondent on behalf of the Tomacan factory the challenges are numerous some of them mentioned were farmers' insincerity, competition with open marketers, non-industrial tomatoes often produced by farmers among others.
- **SWOT Analysis**

  - **Strengths**: The company equipment, land, farms and personnel are strengths of the processors.
  - **Weakness**: The inability of the company to develop its own varieties for production is a big weakness of the company. Also, the inability to offer higher prices to farmers for their produce has been identified as a weakness of the company.
  - **Opportunities**: The availability of Council for Scientific and Industrial Research (SCIR) and the District Agriculture Development Unit (DADU) are considered opportunities for the factory to be revamped in its operations.

- **Threats**: The Company faces serious threats from competition for raw tomatoes with the marketers. Climate change is also a threat to the factory as less or too much rainfall can destroy farms. Farmers' interest to continue to produce non-industrial varieties is a major threat to company to outsource raw materials.

- **Potential activities to**:
  - **Sustain strengths**: According to the production technician/advisor the company still maintains a number of its employees, keeps its equipment and acquires more land for future expansion works.
  - **Address weaknesses**: The Company is working with SCIR and DADU to come out with suitable industrial varieties.
  - **Minimize threats**: The Company has set up a company farm to produce to feed the factory and to engage only trust-worthy farmers on contract production.
  - **Take advantage of opportunities**: The presence of large market for processed tomatoes is an opportunity for the production of tomato paste the technician said. Also the presences of SCIR, DADU, tomato producers to be engaged in contract farming are all opportunities for the company.

8.2.6 Consumers

- **Consumer types**: The types of consumers in the Wenchi District as identified by the research team were individuals and institutional consumer (Wenchi Senior High School, chop bars and food vendors) and they were both interviewed.
- **Product requirements**: The research revealed that both individuals and institutional consumers require both fresh and processed tomatoes. These tomatoes are used for different purposes, for instance, some consumers prefer canned (tin) tomatoes for stew as they said the canned tomato paste taste better and have good flavour than the fresh tomatoes. The fresh tomatoes are however used for soup and 'hot pepper stew'. Some of the tin tomato brands mentioned by consumers were Tasty Tom, Gino Tomato Paste, Pormo, Salsa, etc.

- **Preferences for**: Some of the consumers prefer foreign imported tomatoes both fresh and canned, they said the imported fresh tomatoes are not watery and have a lot of pulp or flesh. Other consumers prefer the locally produced fresh tomatoes with their reason being that they are less expensive. According to the consumers there are not aware of any tomatoes being processed (canned) in Ghana, therefore all the tomato paste they used are believed to be imported. They said should they get made in Ghana tomato paste that is relatively cheaper and as good as the imported ones they will prefer them to the imported ones. There were other consumers who prefer the fresh tomatoes whether imported or locally produced to the canned tomatoes because they believe the canned tomatoes contain chemicals like additives and preservatives that may be injurious to their health. The consumers however indicated that some of the fresh tomatoes contain some amount of chemicals that are applied to control pests and diseases at the point of production and for quick ripening. They said tomatoes that are ripened by chemicals addition have greenish colour inside when cut open.
- **Willingness to pay for pay premium prices**: Some of the consumers said prices are already high so paying for more prices will be difficult for them but it is not to say that they do not want good and quality products they retorted. Some of the consumers said they will not mind paying little more for good quality tomato products.
- **Product**:

  - **Availability**: Consumers said both fresh and tin tomatoes are available throughout the year in the market. However, getting to the latter part of the year, fresh tomatoes become scares in the market a consumer said.
  - **Accessibility**: They said both products (canned and fresh tomatoes) are also accessible and to access the tomatoes one only needs to go to the market, at the market one can buy the quantity needed. Consumers revealed that during the peak of harvest they consume more fresh tomatoes than the canned tomatoes because the fresh tomatoes are cheaper and are found almost everywhere. According to Mrs. Margret Addae, the Senior Domestic Bursar (Matron) of Wenchi Senior High School they buy their tomatoes from the middle-market women

(wholesalers and open-air market retailers) and not directly from the producers. When she was asked why they do not buy directly from the producer she said the producers are those who do not want to sell directly to them because they fear losing their customers.
- **Affordability**: According to the consumers tin tomatoes are more affordable but fresh tomatoes are affordable during the rainy seasons. Mr. John Ndadefo who was one of the consumers said he consumes more of fresh tomatoes during the peak of production due to their lower prices he however switches to the tin (canned) tomato paste during the lean production season when prices of fresh tomatoes are higher.

- **Payment of taxes and levies (if any)**: Consumers indicated that they do not pay levies or taxes for consuming tomatoes whether it is fresh or canned. They however said they believe they pay indirect taxes pass on to them by marketers and processors for consuming the tomatoes.
- **Intra and Interrelationships with**:
  - **Other consumers**: Consumers do not see any relationship that exists between them just because they consume tomatoes they retorted. However, the matron of Wenchi Senior High School indicated that during matrons meetings they inform each other about new products, prices, etc., in the market especially on the tomatoes.
  - **Producers**: Consumers really do not buy directly from producers and thus less relationship exists between them and the producers.
  - **Marketers**: Consumers indicated that they have a business relationship and interact cordially with marketers especially at the point of selling and buying.
  - **Transporters**: Consumers have no direct relationship with the transporters except sometimes for the case of the institutional consumers like the Wenchi Senior High School who go for public transport to transport their foodstuffs including tomatoes, at that point they establish a good and cordial relationship with the transporters.

- **SWOT Analysis**:
  - **Key strengths**: One of the strengths of the consumers is that they have the power to decide on the type of tomatoes to buy and whom to buy from. Consumers also have the ability to bargain for good prices and quality goods. Also, Senior High Schools have available lands and the skilled personnel constituting both the masters and the students to produce their own tomatoes.
  - **Key weaknesses**: Consumers said consuming tomatoes has become a necessity therefore they are forced to buy them either fresh or canned even when prices are higher. Consumers, especially institutional consumers consider their inability to test for chemicals in tomatoes in their laboratories as a weakness.
  - **Key Opportunities**: The availability of many tomato marketers is an opportunity for them as they can choose from a lot of marketers.
  - **Key threats**: The overuse of chemicals on tomatoes is a threat to their lives consumers said.

- **Potential Activities to**:
  - **Sustain strengths**: In order to sustain their strengths, consumers constantly check for damage and unwholesome tomatoes.
  - **Address weaknesses**: Though consumers said they cannot do without tomatoes in their daily meals they mentioned that they sometimes switch to alternative vegetables like garden eggs when prices of tomatoes are very high. Consumers said they address their weaknesses by refusing to buy suspected unwholesome tomatoes.
  - **Minimize threats**: Most of the consumers said they wash fresh tomatoes with clean water however the matron of the Wenchi Senior High School said they wash fresh tomatoes with brine (water containing salt) to reduce to the risk of food poisoning. They said nothing could however be done to canned tomatoes except that they only look out for damaged cans and expiring dates.

- **Take Advantage of the Opportunities**: Consumers could not tell the existence of any opportunity to take advantage of except for the advantage of choosing between fresh and canned tomatoes and, among sellers. Also, they take off the presence of many marketers to look out for good quality and less expensive tomato products.

References

Akudugu, A. M. (2011). Rural banks' financial capital and livelihoods development of women farmers in Ghana. *Journal of Enterprising Communities: People and Places in the Global Economy, 5*(4), 248–264.

Banchirigah, S. M. (2006). How have reforms fuelled the expansion of artisanal mining? Evidence from Sub-Saharan Africa. *Resources Policy, 31*(3), 165–171.

Banchirigah, S. M. (2008). Challenges with eradicating illegal mining in Ghana: A perspective from the grassroots. *Resources Policy, 33*(1), 29–38.

Block, J. R. (1994). A new view of agricultural productivity in Sub-Saharan Africa. *American Journal of Agricultural Economics, 76*(4), 619–624.

Burkhard, B., & Maes, J. (2017). *Mapping ecosystem services* (p. 374). Sofia: Pensoft Publishers.

Costanza, R., de Groot, R., Braat, L., Kubiszeewski, I., Fioramonti, L., Sutton, P., … Grasso, M. (2017). Twenty years of ecosystem services: How far have we come and how far do we still need to go? *Ecosystem Services, 28,* 1–16.

Dewbre, J., & Borot de Battisti, A. (2008). *Agricultural progress in Cameroon, Ghana and Mali: Why it happened and how to sustain it* (OECD Food, Agriculture and Fisheries Working Papers 9), 60.

Diao, X., Hazell, P., Resnick, D., & Thurlow, J. (2007). *The role of agriculture in development: Implications for Sub-Saharn Africa.* Retrieved from Washington, DC.

Díaz, S., Pascual, U., Stenseke, M., Martín-López, B., Watson, R. T., Molnár, Z., … Shirayama, Y., (2018). Assessing nature's contributions to people. *Science, 359*(6373), 270–272.

A.-R. Alhassan and M. Abunga Akudugu,
Supply Chain Management in African Agriculture,
https://doi.org/10.1007/978-3-030-54209-2

Garcia-Winder, M., Riveros, H., Pavez, I., Rodriguez, D., Lam, L., Arias, J., & Herrera, D. (2009). Agrifood chains: A tool for strengthening the institutional framework of the agricultural and rural sector. *Perspectives*, 26–39.

Fan, S., Omilola, B., & Lambert, M. (2009). *Public spending for agriculture in Africa: Trends and composition* (ReSAKSS Working Paper, 28), 20.

FAO. (2000). *The state of food and agriculture: Lessons from the past 50 years.* Rome, Italy.

Frisvold, G., & Ingram, K. (1995). Sources of agricultural productivity growth and stagnation in Sub-Saharan Africa. *Agricultural Economics, 13,* 51–61.

Fulbright, N. R. (2019). *Climate change and the food and Agribusiness sectors: Adaptation, mitigation and innovation.*

Fulginiti, L. E., Perrin, R. K., & Yu, B. (2004). Institutions and agricultural productivity in Sub-Saharan Africa. *Agricultural Economics, 31,* 169–180.

Herrera, D. (2004). *El IICA y las cadenas alimentarias: avances y tareas pendientes.* Discussion 3.

Hilson, G. (2002). The environmental impact of small-scale gold mining in Ghana: Identifying problems and possible solution. *The Geographical Journal, 168*(1), 57–72.

Huge, J., Rochette, A. J., Bethune, S., Parra Paiton, C. C., Vanderhaegen, K., Vandervelden, T.,... Janssens de Bi Bisthoven, L. (2020). *Ecosystem Services,* 42 (2020) 101079.

IPBES. (2019). *Assessment report on pollinators, pollination and food production.* https://ipbs.net/assessment-reports/pollinator.

InterAcademy Council. (2004). *Realizing the promise and potential of African agriculture: Science and technology strategies for improving food security and agricultural productivity in Africa.* Amsterdam, The Netherlands: Royal Netherlands Academy of Arts and Sciences.

Inter-American Institute for Cooperation on Agriculture. (2005). *Policies for rural prosperity: Discussion document.* San José, CR.

Kuah, A. T. H. (2002). Cluster theory and practice: Advantages for the small business locating in a vibrant cluster. *Marketing and Entrepreneurship, 4,* 206.

Lusigi, A., & Thirtle, C. (1997). Total factor productivity and the effects of R&D in African agriculture. *Journal International Development, 9,* 529–538.

Porter, M. (1998). Clusters and the new economics of competition. *Harvard Business Review, 76,* 77–90.

Porter, M. (2008). *Clusters, innovation, and competitiveness: New findings and implications for policy*, Stockholm.

Rao, D. S. P., & Coelli, T. J. (1998). *Catch-up and convergence in global agricultural productivity 1980–1995.* Department of Econometrics, University of New England, Armidale.

Red SICTA (Agricultural Innovation Network Project, Central American Agricultural Technology Integration System); IICA (Inter-American Institute for

Cooperation on Agriculture); COSUDE (Swiss Agency for Development and Cooperation). (2007). *Mapping the white corn and bean agrifood chains in central America*. Managua, NI.

Suhariyanto, K., Lusigi, A., & Thirtle, C. (2001). *Productivity growth and convergence in Asian and African agriculture*. New York: Palgrave.

Tasser, E., Schirpke, U., Zoderer, B. M., & Tappeiner, U., (2020). Towards an integrative assessment of land-use type values from perspective of ecosystem services. *Ecosystem Services*, 42 (2020), 101082.

The Economist Intelligence Unit. (2011). *Fostering innovation-led clusters: A review of leading global practices*, London.

Thirtle, C., Hadley, D., & Townsend, R. (1995). Policy-induced innovation in Sub-Saharan African agriculture: A multilateral Malmquist productivity index approach. *Development Policy Review, 13,* 323–348.

Walter, P. (1990). Neither market nor hierarchy: Network forms of organization, *Research in Organizational Behavior*, 295–336.

World Development Report. (2008). *United nations development programme*. New York.

World Economic Forum. (2019). *System initiative on shaping the future of food. Innovation with a purpose: Improving traceability in Food value chains through technology innovation*.

Index

A.-R. Alhassan and M. Abunga Akudugu,
Supply Chain Management in African Agriculture,
https://doi.org/10.1007/978-3-030-54209-2

The manufacturer's authorised representative in the EU is Springer Nature Customer Service Centre GmbH, Europaplatz 3, 69115 Heidelberg, Germany. If you have any concerns regarding our products, please contact ProductSafety@springernature.com

Printed and bound by CPI Group (UK) Ltd, Croydon, CR0 4YY
07/10/2026
02229221-0001